Selected Titles in This Series

701 **Paul Selick and Jie Wu,** On natural coalgebra decompositions of tensor algebras and loop suspensions, 2000

700 **Vicente Cortés,** A new construction of homogeneous quaternionic manifolds and related geometric structures, 2000

699 **Alexander Fel′shtyn,** Dynamical zeta functions, Nielsen theory and Reidemeister torsion, 2000

698 **Andrew R. Kustin,** Complexes associated to two vectors and a rectangular matrix, 2000

697 **Deguang Han and David R. Larson,** Frames, bases and group representations, 2000

696 **Donald J. Estep, Mats G. Larson, and Roy D. Williams,** Estimating the error of numerical solutions of systems of reaction-diffusion equations, 2000

695 **Vitaly Bergelson and Randall McCutcheon,** An ergodic IP polynomial Szemerédi theorem, 2000

694 **Alberto Bressan, Graziano Crasta, and Benedetto Piccoli,** Well-posedness of the Cauchy problem for $n \times n$ systems of conservation laws, 2000

693 **Doug Pickrell,** Invariant measures for unitary groups associated to Kac-Moody Lie algebras, 2000

692 **Mara D. Neusel,** Inverse invariant theory and Steenrod operations, 2000

691 **Bruce Hughes and Stratos Prassidis,** Control and relaxation over the circle, 2000

690 **Robert Rumely, Chi Fong Lau, and Robert Varley,** Existence of the sectional capacity, 2000

689 **M. A. Dickmann and F. Miraglia,** Special groups: Boolean-theoretic methods in the theory of quadratic forms, 2000

688 **Piotr Hajłasz and Pekka Koskela,** Sobolev met Poincaré, 2000

687 **Guy David and Stephen Semmes,** Uniform rectifiability and quasiminimizing sets of arbitrary codimension, 2000

686 **L. Gaunce Lewis, Jr.,** Splitting theorems for certain equivariant spectra, 2000

685 **Jean-Luc Joly, Guy Metivier, and Jeffrey Rauch,** Caustics for dissipative semilinear oscillations, 2000

684 **Harvey I. Blau, Bangteng Xu, Z. Arad, E. Fisman, V. Miloslavsky, and M. Muzychuk,** Homogeneous integral table algebras of degree three: A trilogy, 2000

683 **Serge Bouc,** Non-additive exact functors and tensor induction for Mackey functors, 2000

682 **Martin Majewski,** ational homotopical models and uniqueness, 2000

681 **David P. Blecher, Paul S. Muhly, and Vern I. Paulsen,** Categories of operator modules (Morita equivalence and projective modules, 2000

680 **Joachim Zacharias,** Continuous tensor products and Arveson's spectral C^*-algebras, 2000

679 **Y. A. Abramovich and A. K. Kitover,** Inverses of disjointness preserving operators, 2000

678 **Wilhelm Stannat,** The theory of generalized Dirichlet forms and its applications in analysis and stochastics, 1999

677 **Volodymyr V. Lyubashenko,** Squared Hopf algebras, 1999

676 **S. Strelitz,** Asymptotics for solutions of linear differential equations having turning points with applications, 1999

675 **Michael B. Marcus and Jay Rosen,** Renormalized self-intersection local times and Wick power chaos processes, 1999

674 **R. Lawther and D. M. Testerman,** A_1 subgroups of exceptional algebraic groups, 1999

For a complete list of titles in this series, visit the
AMS Bookstore at **www.ams.org/bookstore/**.

On Natural Coalgebra Decompositions of Tensor Algebras and Loop Suspensions

MEMOIRS
of the
American Mathematical Society

Number 701

On Natural Coalgebra Decompositions of Tensor Algebras and Loop Suspensions

Paul Selick
Jie Wu

November 2000 • Volume 148 • Number 701 (first of 5 numbers) • ISSN 0065-9266

American Mathematical Society
Providence, Rhode Island

2000 *Mathematics Subject Classification.*
Primary 55P35, 55P45, 20C30, 16W30, 17B01; Secondary 17B50, 17B70.

Library of Congress Cataloging-in-Publication Data

Selick, Paul, 1950–
On natural coalgebra decompositions of tensor algebras and loop suspensions / Paul Selick, Jie Wu
p. cm. — (Memoirs of the American Mathematical Society, ISSN 0065-9266 ; no. 701)
Includes bibliographical references.
ISBN 0-8218-2110-5 (alk. paper)
1. Loop spaces. 2. H-spaces. 3. Representations of groups. I.Wu, Jie, 1964– II. Title. III. Series.
QA3.A57 no. 701
[QA612.76]
510 s—dc21
[514′.24] 00-059369

Memoirs of the American Mathematical Society

This journal is devoted entirely to research in pure and applied mathematics.

Subscription information. The 2000 subscription begins with volume 143 and consists of six mailings, each containing one or more numbers. Subscription prices for 2000 are $466 list, $419 institutional member. A late charge of 10% of the subscription price will be imposed on orders received from nonmembers after January 1 of the subscription year. Subscribers outside the United States and India must pay a postage surcharge of $30; subscribers in India must pay a postage surcharge of $43. Expedited delivery to destinations in North America $35; elsewhere $130. Each number may be ordered separately; *please specify number* when ordering an individual number. For prices and titles of recently released numbers, see the New Publications sections of the *Notices of the American Mathematical Society.*

Back number information. For back issues see the *AMS Catalog of Publications.*

Subscriptions and orders should be addressed to the American Mathematical Society, P. O. Box 845904, Boston, MA 02284-5904. *All orders must be accompanied by payment.* Other correspondence should be addressed to Box 6248, Providence, RI 02940-6248.

Memoirs of the American Mathematical Society is published bimonthly (each volume consisting usually of more than one number) by the American Mathematical Society at 201 Charles Street, Providence, RI 02904-2294. Periodicals postage paid at Providence, RI. Postmaster: Send address changes to Memoirs, American Mathematical Society, P. O. Box 6248, Providence, RI 02940-6248.

This publication is indexed in *Science Citation Index*®, *SciSearch*®, *Research Alert*®, *CompuMath Citation Index*®, *Current Contents*®/*Physical, Chemical & Earth Sciences.*
Printed in the United States of America.

♾ The paper used in this book is acid-free and falls within the guidelines established to ensure permanence and durability.
Visit the AMS home page at URL: `http://www.ams.org/`

10 9 8 7 6 5 4 3 2 1 05 04 03 02 01 00

Contents

ABSTRACT. We consider functorial decompositions of $\Omega\Sigma X$ in the case where X is a p-torsion suspension. By means of a geometric realization theorem, we show that the problem can be reduced to the one obtained by applying homology: that of finding natural coalgebra decompositions of tensor algebras. We solve the algebraic problem and give properties of the piece $A^{\text{min}}(V)$ of the decomposition of $T(V)$ which contains V itself, including verification of the Cohen conjecture that in characteristic p the primitives of $A^{\text{min}}(V)$ are concentrated in degrees of the form p^t. The results tie in with the representation theory of the symmetric group and in particular produce the maximum projective submodule of the important S_n-module Lie(n).

1. INTRODUCTION

In classifying any mathematical structure, it is helpful to analyze the irreducible or indecomposable components. The geometric problem studied in this paper is that of finding natural homotopy decompositions of the loop-suspension $\Omega\Sigma X$ where X is itself a p-torsion suspension for some prime p. The first main result is the construction of a functor $A(X)$ which gives the smallest natural homotopy retract of $\Omega\Sigma X$ whose mod p homology contains $H_*(X;\mathbb{Z}/p\mathbb{Z})$. (See Theorem 1.1). Among the properties of $A(X)$ which we show is that, as conjectured by Cohen, the primitives in the coalgebra $H_*\big(A(X);\mathbb{Z}/p\mathbb{Z}\big)$ are concentrated in weights of the form p^t within the tensor algebra $T\big(\bar{H}(X;\mathbb{Z}/p\mathbb{Z})\big) \cong H_*(\Omega\Sigma X;\mathbb{Z}/p\mathbb{Z})$.

By means of the "Geometric Realization Theorem" (see Theorem 1.3) we show the above geometric problem to be equivalent to the algebraic problem of finding natural coalgebra decompositions of (ungraded) primitively generated tensor algebras over the field $\mathbb{F}_p$. With the problem thus reduced to algebra we consider the latter problem over an arbitrary field (not necessarily of characteristic p). The solution to this problem (and thus to the original geometric problem) is given in Theorem 1.5 and its generalization Theorem 6.5, which gives the more complete decomposition.

We refer to Theorem 6.5 as a functorial version of the Poincaré-Birkhoff-Witt Theorem. Recall that the Poincaré-Birkhoff-Witt Theorem gives a coalgebra isomorphism $U(L) \cong S(L)$ for a Lie algebra L, where $S(W)$ denotes the free commutative algebra on W, and the coalgebra structure on $S(W)$ is the one under which it becomes a primitively generated Hopf algebra. (That is, elements of W are primitive.) Applying this in the case where $L = L(V)$, the free Lie algebra on the vector space V, gives a coalgebra decomposition $PBW_V : T(V) = U\big(L(V)\big) \cong S\big(L(V)\big) \cong S\big(\oplus_{n=1}^{\infty} L_n(V)\big) \cong \otimes_{n=1}^{\infty} S\big(L_n(V)\big)$. However this isomorphism is **not** natural with respect to maps of the vector space V; it is natural with respect to maps of **ordered** bases for vector spaces. For example, if $V = \langle x, y\rangle$ with $x < y$ then $PBW_V(x \otimes y) = xy$ while $PBW_V(y \otimes x) = xy - [x,y]$ so PBW_V does not commute with the map which interchanges x and y. Without concerning ourselves for the moment as to the definition of the factors $A^{\min}(V; L_n^{\max})$ which appear in Theorem 6.5, we observe that if $\operatorname{char}(\mathbf{k}) = 0$ then according to Proposition 6.12, $A^{\min}(V; L_n^{\max}) \cong S\big(L_n(V)\big)$, so that the right hand side of Theorem 6.5 is identical to that of the Poincaré-Birkhoff-Witt decomposition in this case. However the isomorphism is different. For example, if ϕ_V is the isomorphism of Theorem 6.5, then $\phi_V(x \otimes y) = (xy + [x,y])/2$ and $\phi_V(y \otimes x) = (xy - [x,y])/2$ and unlike PBW_V, this is indeed natural with respect to maps of V.

As the reader has no doubt already guessed upon seeing the division by 2 in the preceding formulas, Theorem 6.5 becomes substantially different when $\operatorname{char}\mathbf{k} > 0$.

The authors are grateful to NSERC and the Fields Institute for support.
Received by the editor August 12, 1998.

Since the maps are required to commute with interchange of variables, the representation theory over $\mathbf{k}$ of the symmetric groups S_n must be involved. In fact, any natural self-transformation of the tensor algebra functor $T(\ )$ is determined by a sequence of elements $(\lambda_n) \in \mathbf{k}(S_n)$ from the group algebras (see Lemma 2.1). Let $\bar{V} = \langle x_1, x_2, \dots, x_n\rangle$ be an n-dimensional vector space over $\mathbf{k}$. The $\mathbf{k}(S_n)$-module $\mathrm{Lie}(n)$ is defined as the (vector) subspace of $\bar{V}^{\otimes n} \subset T(\bar{V})$ generated by the iterated commutators $\{[\dots[[x_{\sigma(1)}, x_{\sigma(2)}], x_{\sigma(3)}], \dots, x_{\sigma(n)}]\}_{\sigma\in S_n}$, made into an $\mathbf{k}(S_n)$-module by letting S_n act by permutation on the basis of $\bar{V}$. This intriguing module has arisen in recent work of Cohen, Dwyer-Hirschorn, and others, because it appears in the homology of various interesting spaces. In spite of much effort, representation theory of the symmetric groups in the modular case (when the characteristic of the field divides the order of the group) remains, for the most part, a mystery. In particular, not much is known about the algebraic properties of the module $\mathrm{Lie}(n)$. An important aspect of our work is that it sheds some light on these properties as follows. Our natural decomposition of $T(\ )$ yields a decomposition of $\mathbf{k}(S_n)$ for each n resulting in the construction of an important projective $\mathbf{k}(S_n)$-submodule of $\mathrm{Lie}(n)$, which we call $\mathrm{Lie}^{\max}(n)$. We show (Theorem 7.4) that it is the maximum projective $\mathbf{k}(S_n)$-submodule of $\mathrm{Lie}(n)$ in the sense that any projective $\mathbf{k}(S_n)$-submodule of $\mathrm{Lie}(n)$ is a $\mathbf{k}(S_n)$-retract of $\mathrm{Lie}^{\max}(n)$. The interest here is in the modular case, since it is a well known consequence of the Dynkin-Specht-Weber relation $\beta_n \circ \beta_n = n\beta_n$ that if $\operatorname{char}\mathbf{k}$ does not divide n then $\mathrm{Lie}(n) = \operatorname{Im}\beta_n$ is itself projective (and thus $\mathrm{Lie}^{\max}(n) = \mathrm{Lie}(n)$ for such n). Here $\beta_n : V^{\otimes n} \to V^{\otimes n}$ is given by $\beta(v_1 v_2 v_3 \cdots v_n) = [\dots[[v_1, v_2], v_3], \dots v_n]$.

Section 8 of this paper considers the decomposition $T(V) \cong A(V) \otimes B(V)$ of Theorem 1.5 and describes properties of the functor $A(\ )$ (denoted also as $A^{\min}(\ )$). This section contains the proof of the Cohen conjecture (Theorem 8.3) referred to earlier. Any decomposition of a Hopf algebra determines Hopf algebra structures on each of the factors, but in general these have no good Hopf algebra properties, and in particular the inclusion and projection maps onto the factors need not be Hopf algebra maps. Since part of the content of Theorem 1.4 is that this decomposition has the extra property that $B(V)$ is a sub Hopf algebra of $T(V)$, we are able to use our solution to the Cohen conjecture to give an upper bound on the size of $A(V)$ (Corollary 8.5). Although we give (Corollary 8.27) a complete description of $A_n^{\min}(V)$ in the first interesting case ($n = p$), we are unable to give a complete description in general. It turns out that $A_n^{\min}(V)$ agrees with the upper bound given by Corollary 8.5 when $n < p^2$, however we know (see subsection 11.2) that equality fails for $n = p^2$ when $p = 3$. This section also contains a coalgebra filtration on $A_n^{\min}(V)$ which is used to give a lower bound on its growth.

The penultimate section of this paper contains a second coalgebra decomposition of $T(V)$ (Theorem 10.7), this time decomposing the factor $B(V)$ into tensor algebras rather than into minimal coalgebra factors. This leads to the construction of another projective $\mathbf{k}(S_n)$-submodule of $\mathrm{Lie}(n)$ which we denote $\mathrm{Lie}'(n)$. $\mathrm{Lie}'(n)$ appears, in some cases, to be more readily computable than $\mathrm{Lie}^{\max}(n)$, and serves as a lower bound.

While, by construction, there are no nontrivial natural retracts of $A(\ )$, it is possible that $A(X)$ decomposes further for particular spaces X. Indeed this is known to happen in the case where X is the Moore space $P^n(p) = S^{n-1} \cup_p e^n$ for $p > 2$, according to the complete decomposition of these spaces given by Cohen-Moore-Neisendorfer ([5, 6, 7, 14]). (See subsection 11.5 for details on where our decomposition in this case differs from the Cohen-Moore-Neisendorfer decomposition.) There is no corresponding decomposition known however when $p = 2$, and one of the primary motivations for considering the problems discussed in this paper is the construction of the space $A\big(P^n(2)\big)$. It is hoped that its study will lead to greater knowledge of the homotopy theory of the mod 2 Moore space and perhaps give information about mod 2 exponents for the homotopy groups of spheres.

The final section collects a number of examples, some of which were referred to above. It contains calculations of some of the functors for low values of n, for spaces with only two cells, and some discussion of how our functorial Poincaré-Birkhoff-Witt isomorphism compares with other maps which are in the literature in characteristic 0.

As noted above, the algebraic results of this paper are also valid in characteristic 0, however the resulting topological theorems are already well known in that case. Since a rational co-H-space is a wedge of spheres, the rational analogue of the geometric problem is not an interesting question. We also note that the condition that X be a suspension can be weakened somewhat. The proofs go through without modification in the case where X is a simply connected co-H-space, and all that is really needed is that X be simply connected and conilpotent. (That is, that the reduced diagonal given by the composite $X \to X \times X -> X \wedge X$ be null.)

The remainder of this introduction describes the results in more detail.

Let X be a p-torsion suspension and let $\Omega\Sigma X$ be the loop space of the suspension of X. The unreduced and reduced mod p homology of X will be denoted by $H_*(X)$ and $\bar{H}_*(X)$, respectively. Recall that $H_*(\Omega\Sigma(X))$ is isomorphic to the tensor algebra $T(V)$ as Hopf algebras, where $V = \bar{H}_*(X)$ and $T(V)$ is given the Hopf algebra structure determined by making V primitive. Let $L_n(V)$ be the set of homogeneous Lie elements of tensor length n in the tensor algebra $T(V)$ for any (ungraded or connected graded) vector space V. In the following theorem, $V = \bar{H}_*(X)$.

Theorem 1.1. *Let X be a p-torsion suspension. Then there is a natural homotopy decomposition*

$$\Omega\Sigma X \simeq A(X) \times B(X)$$

such that

1) $V \subseteq H_*(A(X))$;
2) $B(X)$ *is a loop suspension and the injection* $B(X) \to \Omega\Sigma X$ *is a loop map;*
3) $L_n(V) \subseteq H_*(B(X))$ *if* n *is not a power of* p.

Remark 1.2. *The functor A is the geometric realization of the algebraic functor $A^{\min}$ described later which gives natural minimal coalgebra retracts of tensor algebras. The space $B(X)$ is the loop suspension of the wedge of certain functorial retracts of the self smash product $X^{(n)}$ for $n \geq 2$. So this theorem can be regarded as a functorial version (on general loop suspensions) of the classical theorems on the decomposition of $\Omega P^n(p^r)$* [5, Theorem 1.1].

If there is no possible confusion, we use the notation V for either a graded module or an ungraded module. Otherwise, we will write V^u for ungraded modules. Also we use the convention that an ungraded module V^u can be regarded as a connected graded module by assigning dimension 2 to elements of V^u. [12].

The proof of this theorem will be given through the following steps. The first step is to reduce the problem of natural decompositions of loop spaces of double suspensions to the problem of natural coalgebra decomposition of tensor algebras.

Theorem 1.3 (Geometric Realization Theorem). *Let $f_{V^u}\colon T(V^u) \to T(V^u)$ be a morphism of ungraded coalgebras over $\mathbb{Z}/p\mathbb{Z}$ such that f_{V^u} is a natural transformation of the functor T. Then there is a functorial morphism of graded coalgebras $f^{\mathrm{grade}}\colon T(V) \to T(V)$ for any connected graded module V such that*

(1). $f^{\mathrm{grade}}_{V^u} = f_{V^u}$ *for any ungraded module* V^u;
(2). $f^{\mathrm{grade}}_{V}\colon T(V) \to T(V)$ *is a functorial morphism of bigraded coalgebras, where the bi-grading in $T(V)$ is given in the canonical way;*
(3). *for any suspension X, then there exists a map $\phi_X\colon \Omega\Sigma X \to \Omega\Sigma X$, "functorial up to homotopy" in X, such that*

$$\phi_{X*} = f^{\mathrm{grade}}_{\bar{H}_*(X)}\colon H_*(\Omega\Sigma X) = T(\bar{H}_*(X)) \to T(\bar{H}_*(X)).$$

The map $\phi_X\colon \Omega\Sigma X \to \Omega\Sigma X$ is functorial up to homotopy means that for any suspensions X and Y and any map $f\colon X \to Y$, where we do not assume that f is a suspension, the diagram

$$\begin{array}{ccc} \Omega\Sigma X & \xrightarrow{\phi_X} & \Omega\Sigma X \\ \Big\downarrow{\scriptstyle \Omega\Sigma f} & & \Big\downarrow{\scriptstyle \Omega\Sigma f} \\ \Omega\Sigma Y & \xrightarrow{\phi_Y} & \Omega\Sigma Y \end{array}$$

commutes up to homotopy. The idea of the proof of this theorem is to compare the Cohen progroup [3, 4] for natural transformations of loop spaces of double suspensions with the progroup introduced in Section 2.

The second step is to show the existence of functorial indecomposable coalgebra retracts of tensor algebras. In the following theorems, V means an ungraded **k**-module and the ground ring **k** is any field. Let M be a functor from **k**-modules to **k**-modules such that $M(V)$ is a sub **k**-module of $T(V)$. A sub coalgebra $A(V)$ of $T(V)$ is called a *natural minimal coalgebra retract over* $M(V)$ if

1) A is a functor and M is a subfunctor of A;
2) A is a retract of T as functors from **k**-modules to coalgebras;
3) $A(V)$ satisfies the following minimality condition:
   if $C(V)$ is a sub coalgebra of $T(V)$ that satisfies conditions 1) and 2) above, then A is a retract of C as functors from **k**-modules to coalgebras.

Theorem 1.4 (Existence of natural minimal coalgebra retracts). *For any functor M from **k**-modules to **k**-modules such that $M(V)$ is a sub **k**-module of $T(V)$, there exists a unique, up to isomorphism, natural minimal coalgebra retract of $T(V)$ over $M(V)$.*

The natural minimal coalgebra retract of $T(V)$ over V will be denoted $A^{\min}(V)$. The proof of this theorem will be given by abstract argument instead of explicit determination.

The third step is to describe some properties of the primitive elements of $A^{\min}(V)$.

Theorem 1.5. *There is a natural sub Hopf algebra of $T(V)$, which is denoted by $B^{\max}(V)$, such that*

1) *there is a natural coalgebra decomposition*
$$T(V) \cong B^{\max}(V) \otimes A^{\min}(V);$$
2) $L_n(V) \subseteq B^{\max}(V)$ *if n is not a power of p, where p is the characteristic of **k**.*

Note that while any coalgebra decomposition of a Hopf algebra determines a Hopf algebra structure on each of the factors, these structures are not in general compatible.

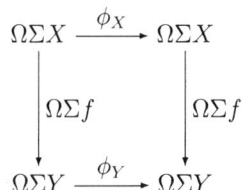

In this case however, as noted in the theorem, the inclusion $B^{\max}(V) \hookrightarrow T(V)$ is multiplicative, although the inclusion of $A^{\min}(V)$ and projections onto the factors are not. Similarly, in the geometric realization, while each factor becomes an H-space, $B(X) \to \Omega\Sigma X$ is an H-map, but the other maps are not.

The functorial retract $A(X)$ in Theorem 1.1 is given by the geometric realization of the functor $A^{\min}$ as follows. Let $f\colon T(V) \to T(V)$ be the composite

$$T(V) \xrightarrow{r_{A^{\min}}} A^{\min}(V) \hookrightarrow T(V),$$

where $r_{A^{\min}}$ is a functorial coalgebra retraction. Given a suspension X, let $\phi_X\colon \Omega\Sigma X \to \Omega\Sigma X$ be the geometric realization of f as in Theorem 1.3. Let $A^{\min}(X)$ be the homotopy colimit

$$A^{\min}(X) = \mathrm{hocolim}_{\phi_X} \Omega\Sigma X.$$

Then the functorial retract $A(X)$ is $A^{\min}(X)$. Similarly, the complementary factor $B(X)$ is given by the geometric realization of the functor $B^{\max}$.

The retract $A^{\min}(X)$ catches the maximum exponent of $\pi_*(\Sigma X)$ in the following sense. Let $\exp\bigl(\pi_*(X)\bigr) \leq \infty$ denote the exponent of the homotopy groups of X and let $X^{(n)}$ denote the n-fold self smash product of X.

Theorem 1.6. *Let X be a p-torsion suspension. Then there are inequalities*

$$\exp\bigl(\pi_*(A^{\min}(X))\bigr) \leq \exp\bigl(\pi_*(\Sigma X)\bigr) \leq \max\{\exp\bigl(\pi_*(A^{\min}(X^{(n)}))\bigr); 1 \leq n < \infty\}.$$

The article is organized as follows. In section 2, we study natural coalgebra transformations of tensor algebras. The proof of Theorem 1.3 is given in section 3, where Theorem 1.3 is Corollary 3.5. In section 4, we study the existence of natural minimal coalgebra retracts of tensor algebras. The proof of Theorem 1.4 is given in this section where Theorem 1.4 is Theorem 4.12. We give some lemmas on coalgebras in section 5. In section 6, we study the functorial version of the Poincaré-Birkhoff-Witt theorem. The fact that the complementary factor $B^{\max}(V)$ is a sub Hopf algebra of $T(V)$ is given by Proposition 6.1. This proves the first part of Theorem 1.5. We give some lemmas on the S_n-module $\mathrm{Lie}(n)$ in section 7. In section 8, we study the functors $A^{\min}$ and $B^{\max}$ over a field with non-zero characteristic. Sections second part of Theorem 1.5 is Corollary 8.4. The proofs of Theorems 1.1 and 1.6 are given in section 9. Section 10 discusses the projective $\mathbf{k}(S_n)$-module $\mathrm{Lie}'(n)$ and the decomposition of $B(V)$ in terms of tensor algebras. Section 11 contains examples.

2. Natural coalgebra transformations of tensor algebras

In this section, the ground ring is a field $\mathbf{k}$ and V is a connected graded $\mathbf{k}$-module. The purpose of the section is to collect some information about natural transformations of tensor algebras culminating in Theorem 2.7 and Corollary 2.9. These will be used in the proof the geometric realization thereom in Section 3.

Lemma 2.1. *Let $\phi_V\colon V^{\otimes n} \to V^{\otimes m}$ be a functorial map of graded $\mathbf{k}$-modules and let $a_1, \dots, a_n$ be n homogeneous elements in V.*

1) *If $n = m$, then the element $\phi_V(a_1 \otimes \cdots \otimes a_n)$ lies in to the sub graded $\mathbf{k}$-module of $V^{\otimes n}$ spanned by the elements*

$$a_{\sigma(1)} \otimes \cdots \otimes a_{\sigma(n)},$$

as σ runs through all elements in S_n.

2) *If $n \neq m$, then ϕ_V is the zero map.*

Proof. Let $a_1 \cdots a_n$ denote $a_1 \otimes \cdots \otimes a_n$ in the tensor product. Let $\bar{V}$ be the graded $\mathbf{k}$-module with basis $\{x_1, \dots, x_n\}$ where $|x_j| = |a_j|$ for $1 \leq j \leq n$. Let $f\colon \bar{V} \to V$ be the map of graded $\mathbf{k}$-modules given by $f(x_j) = a_j$. By hypothesis there is a commutative diagram

$$\begin{array}{ccc}
\bar{V}^{\otimes n} & \xrightarrow{f^{\otimes n}} & V^{\otimes n} \\
\Big\downarrow \phi_{\bar{V}} & & \Big\downarrow \phi_V \\
\bar{V}^{\otimes m} & \xrightarrow{f^{\otimes m}} & V^{\otimes m}.
\end{array}$$

It suffices to show that the assertions hold for the case where $V = \bar{V}$ and $a_j = x_j$ for $1 \leq j \leq n$. Let $d_j\colon \bar{V} \to \bar{V}$ be the map of graded $\mathbf{k}$-modules defined by

$$d_j(x_i) = \begin{cases} x_i & \text{if} \quad i \neq j \\ 0 & \text{if} \quad i = j. \end{cases}$$

1) Let d_j denote $d_j^{\otimes n}\colon \bar{V}^{\otimes n} \to \bar{V}^{\otimes n}$ for $1 \leq j \leq n$. Then we have

$$d_j(x_{i_1} \cdots x_{i_n}) = \begin{cases} x_{i_1} \cdots x_{i_n} & \text{if } j \notin \{i_1, \dots, i_n\} \\ 0 & \text{if } j \in \{i_1, \dots, i_n\}. \end{cases}$$

Let $\gamma_n(x_1, \dots, x_n)$ be the sub graded $\mathbf{k}$-module of $V^{\otimes n}$ spanned by the elements

$$x_{\sigma(1)} \cdots x_{\sigma(n)},$$

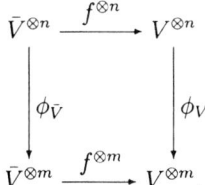

as σ runs through all elements in S_n. Then

$$\gamma_n(x_1,\dots,x_n)\subseteq\bigcap_{1\le j\le n}\mathrm{Ker}(d_j).$$

Conversely, let

$$\alpha=\sum_I k_I x_{i_1}\cdots x_{i_n}\in\bigcap_{1\le j\le n}\mathrm{Ker}(d_j),$$

where $I=(i_1,\dots,i_n)$ with $1\le i_p\le n$ for $1\le p\le n$ and $k_I\in R$. Let $\alpha_1=\sum_{1\notin I}k_I x_{i_1}\cdots x_{i_n}$ and let $\alpha_2=\sum_{1\in I}k_I x_{i_1}\cdots x_{i_n}$. Then

$$\alpha=\alpha_1+\alpha_2,\qquad d_1(\alpha_1)=\alpha_1,\qquad d_1(\alpha_2)=0.$$

Notice that $d_1(\alpha)=0$. Thus

$$\alpha_1=d_1(\alpha_1)=d_1(\alpha)=0$$

and so

$$\alpha=\alpha_2=\sum_{1\in I}k_I x_{i_1}\cdots x_{i_n}.$$

Inductively, we have

$$\alpha=\sum_{1,2,\dots,n\in I}k_I x_{i_1}\cdots x_{i_n}\in\gamma_n(x_1,\dots,x_n).$$

Thus

$$\gamma_n(x_1,\dots,x_n)=\bigcap_{1\le j\le n}\mathrm{Ker}(d_j).$$

Notice that

$$d_j\circ\phi_{\bar V}=\phi_{\bar V}\circ d_j$$

for $1\le j\le n$. Thus

$$\phi_{\bar V}(\gamma_n(x_1,\dots,x_n))=\phi_{\bar V}(\bigcap_{1\le j\le n}\mathrm{Ker}(d_j))\subseteq\bigcap_{1\le j\le n}\mathrm{Ker}(d_j)=\gamma_n(x_1,\dots,x_n).$$

Assertion 1) follows.

2) We consider two cases.

Case I: $m<n$.

From the commutative diagram

$$\begin{array}{ccc} \bar{V}^{\otimes n} & \xrightarrow{\phi_{\bar{V}}} & \bar{V}^{\otimes m} \\ \downarrow d_j^{\otimes n} & & \downarrow d_j^{\otimes m} \\ \bar{V}^{\otimes n} & \xrightarrow{\phi_{\bar{V}}} & \bar{V}^{\otimes m} \end{array}$$

for $1 \leq j \leq n$, one gets

$$\phi_{\bar{V}}(x_1 \cdots x_n) \in \cap_{1\leq j\leq n} Ker(d_j^{\otimes m} \colon V^{\otimes m} \to V^{\otimes m}) = 0$$

by the proof of assertion (1) and Case I of assertion 2) follows.

Case II: $m > n$.

Let V be any connected graded $\mathbf{k}$-module of finite type. Notice that the dual map

$$\phi_V^* \colon (V^*)^{\otimes m} \to (V^*)^{\otimes n}$$

is natural transformation for $\mathbf{k}$-modules V^* of finite type. By Case I of assertion 2), we have

$$\phi_V^* = 0$$

for any connected graded $\mathbf{k}$-module of finite type and Case II of assertion 2) follows.

□

Let V be a connected graded $\mathbf{k}$-module. Then $T(V) = \oplus_{n\geq 0} V^{\otimes n}$ is a bigraded $\mathbf{k}$-module.

Corollary 2.2. *Let $\phi_V \colon T(V) \to T(V)$ be a functorial map of graded $\mathbf{k}$-modules. Then ϕ_V is a functorial map of bi-graded $\mathbf{k}$-modules.*

Definition 2.3. *The* **James (coalgebra) filtration** *of the tensor algebra $T(V)$ is defined by*

$$J_n(V) = \oplus_{0\leq j\leq n} T_j(V)$$

with the canonical comultiplication.

Let C be a (graded) coalgebra and let A be a (graded) algebra. Recall that the convolution product $f * g$ of $f, g \colon C \to A$ is defined by

$$C \xrightarrow{\psi} C \otimes C \xrightarrow{f \otimes g} A \otimes A \xrightarrow{\mu} A,$$

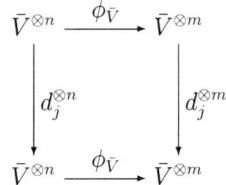

where $\psi \colon C \to C \otimes C$ is the comultiplication and $\mu \colon A \otimes A \to A$ is the multiplication. (See [12].) Let $\mathrm{Hom}_{\mathrm{coalg}}(J_n(-), T(-))$ and $\mathrm{Hom}_{\mathrm{coalg}}(T(-), T(-))$ denote the set of all of natural transformations of coalgebras from $J_n(V)$ and $T(V)$ to $T(V)$, respectively, with the multiplications given by the convolution product. The James filtration

$$J_0(V) \subseteq J_1(V) \subseteq \cdots \subseteq J_n(V) \subseteq \cdots \subseteq T(V)$$

induces a cofiltration of groups

$$\mathrm{Hom}_{\mathrm{coalg}}(T(-),T(-)) \to \cdots \to \mathrm{Hom}_{\mathrm{coalg}}(J_n(-),T(-)) \to \cdots \to \mathrm{Hom}_{\mathrm{coalg}}(J_0(-),T(-))$$

with

$$\mathrm{Hom}_{\mathrm{coalg}}(T(-),T(-)) = \lim_{\substack{\leftarrow \\ n}} \mathrm{Hom}_{\mathrm{coalg}}(J_n(-),T(-)).$$

Let $L_n(V)$ denote the set of homogeneous Lie elements of tensor length n in the tensor algebra $T(V)$ and let $\mathrm{Hom}_{\mathbf{k}}((-)^{\otimes n}, L_n(-))$ denote the set of all of natural transformations of graded modules from the functor $(-)^{\otimes n}$ to the functor L_n.

Proposition 2.4. *Let Γ_n be the kernel of the homomorphism*

$$\mathrm{Hom}_{\mathrm{coalg}}(J_n(-),T(-)) \to \mathrm{Hom}_{\mathrm{coalg}}(J_{n-1}(-),T(-)).$$

Then there is an isomorphism of groups

$$\Gamma_n \xrightarrow{\cong} \mathrm{Hom}_{\mathbf{k}}((-)^{\otimes n}, L_n(-)).$$

Proof. Let $\phi_V \in \Gamma_n$ and let $\theta(\phi_V)\colon V^{\otimes n} \to T(V)$ be the restriction of ϕ to the summand $V^{\otimes n}$ of $J_n(V)$. By Lemma 2.1, we have that $\theta(\phi_V)(V^{\otimes n}) \subseteq V^{\otimes n}$. By hypothesis, ϕ_V is a morphism of graded coalgebras. From the commutative diagram

$$\begin{array}{ccc} T(V) & \xrightarrow{\phi_V} & T(V) \\ \downarrow{\scriptstyle \psi} & & \downarrow{\scriptstyle \psi} \\ T(V)\otimes T(V) & \xrightarrow{\phi_V \otimes \phi_V} & T(V)\otimes T(V) \end{array}$$

together with the condition that ϕ_V belongs to Γ_n, we have

$$\theta(\phi_V)(V^{\otimes n}) \subseteq P_n(V),$$

where $P_n(V)$ is the set of primitive elements of tensor length n in $T(V)$.

Let $a_1, \ldots, a_n$ be homogeneous elements in V and let $\bar{V}$ be the graded $\mathbf{k}$-module with basis $\{x_1, \ldots, x_n\}$ where $|x_j| = |a_j|$ for $1 \leq j \leq n$. Let γ_n be the graded submodule of $\bar{V}^{\otimes n}$ generated by the homogeneous elements

$$x_{\sigma(1)} \cdots x_{\sigma(n)}$$

for $\sigma \in S_n$. By Lemma 2.1, we have that

$$\theta(\phi_{\bar{V}})(x_1 \cdots x_n) \in \gamma_n \cap P_n(\bar{V}) \subseteq L_n(\bar{V}).$$

Let $f\colon \bar{V} \to V$ be a $\mathbf{k}$-linear map given by

$$f(x_j) = a_j$$

for $1 \leq j \leq n$. From the commutative diagram

$$\begin{array}{ccc} \bar{V}^{\otimes n} & \xrightarrow{f^{\otimes n}} & V^{\otimes n} \\ \Big\downarrow{\scriptstyle \theta(\phi_{\bar{V}})} & & \Big\downarrow{\scriptstyle \theta(\phi_V)} \\ T(\bar{V}) & \xrightarrow{T(f)} & T(V), \end{array}$$

we have that $\theta(\phi_V)(a_1 \cdots a_n) \in L_n(V)$ for any $a_1, \ldots, a_n \in V$ and so one gets a function

$$\theta\colon \Gamma_n \to \mathrm{Hom}_{\mathbf{k}}((-)^{\otimes n}, L_n(-)).$$

It is easy to check that

(1). $\theta(\phi_V * \phi'_V) = \theta(\phi_V) + \theta(\phi'_V)$ for $\phi_V, \phi'_V \in \Gamma_n$;
(2). θ is a monomorphism.

Conversely, given a natural $\mathbf{k}$-linear map $\lambda_V\colon V^{\otimes n} \to L_n(V)$, let $\phi_V\colon J_n(V) \to T(V)$ be the map defined by

(1). $\phi_V\colon T_0(V) = \mathbf{k} \to T_0(V) = \mathbf{k}$ is the identity;
(2). $\phi_V\colon T_j(V) \to T_j(V)$ is zero for $0 < j < n$;
(3). $\phi_V = \lambda_V\colon T_n(V) = V^{\otimes n} \to L_n(V) \subseteq T_n(V)$.

Then $\phi_V \in \Gamma_n$ with $\theta(\phi_V) = \lambda_V$. The assertion follows. □

Let $I = (i_1, \ldots, i_n) \in \mathbb{N}^n$ be an n-tuple of positive integers. The length $l(I)$ is defined by

$$l(I) = i_1 + i_2 + \cdots + i_n.$$

Let $\bar{V}(I)$ be the graded $\mathbf{k}$-module wth basis $\{x_1, \ldots, x_n\}$ where $|x_j| = i_j$ for $1 \leq j \leq n$. Let $\mathrm{Lie}^I(n)$ be the graded sub $\mathbf{k}$-module of $\bar{V}(I)^{\otimes n}$ generated by the n-fold graded commutators

$$[[x_{\sigma(1)}, x_{\sigma(2)}], \cdots, x_{\sigma(n)}]$$

for $\sigma \in S_n$.

Lemma 2.5. *There is an isomorphism of groups*

$$\mathrm{Hom}_{\mathbf{k}}((-)^{\otimes n}, L_n(-)) \cong \prod_{1\leq q<\infty} \bigoplus_{l(I)=q} \mathrm{Lie}^I(n).$$

Proof. For a connected graded $\mathbf{k}$-module $V=\bigoplus_{q=1}^{\infty} V_q$, we have natural decomposition

$$V^{\otimes n} \cong \bigoplus_{q=1}^{\infty} \bigoplus_{i_1+\cdots+i_n=q} V_{i_1}\otimes\cdots\otimes V_{i_n}$$

in terms of grading. Thus there is a natural decomposition

$$\mathrm{Hom}_{\mathbf{k}}(V^{\otimes n}, L_n(V)) \cong \prod_{q=1}^{\infty} \bigoplus_{i_1+\cdots+i_n=q} \mathrm{Hom}_{\mathbf{k}}(V_{i_1}\otimes\cdots\otimes V_{i_n}, L_n(V)).$$

The assertion follows from the proof of Proposition 2.4. □

Let $\bar{J}_n(V)=J_n(V)/J_0(V)$. Notice that there is a unique Hopf algebra structure on $T(\bar{J}_n(V))$ such that the comultiplication on $T(\bar{J}_n(V))$ is induced by the comultiplication on $J_n(V)$.

Lemma 2.6. *There is a functorial isomorphism of Hopf algebras*

$$h_n\colon T(\bar{J}_n(V)) \longrightarrow T(\bigoplus_{k=1}^{n} V^{\otimes k}),$$

such that the diagram

$$\begin{array}{ccc} T(\bar{J}_n(V)) & \xrightarrow[\cong]{h_n} & T(\bigoplus_{k=1}^{n} V^{\otimes k}) \\ \uparrow & & \uparrow \\ T(\bar{J}_{n-1}(V)) & \xrightarrow[\cong]{h_{n-1}} & T(\bigoplus_{k=1}^{n-1} V^{\otimes k}) \end{array}$$

commutes, where the Hopf algebra structure on $T(\bigoplus_{k=1}^{n} V^{\otimes k})$ is the one in which the elements in $\bigoplus_{k=1}^{n} V^{\otimes k}$ are primitive.

Proof. Let $H_k\colon T(V)\to T(V^{\otimes k})$ be the James-Hopf map. (See [8].) Let $f_n\colon J_n(V)\to T(\bigoplus_{k=1}^{n} V^{\otimes k})$ be the composite

$$J_n(V) \xrightarrow{\text{comult}} \bigotimes^{n} J_n(V) \hookrightarrow \bigotimes^{n} T(V) \xrightarrow{\otimes_{k=1}^{n} H_k} \bigotimes_{k=1}^{n} T(V^{\otimes k}) \xrightarrow{(i_1,i_2,\ldots,i_n)} T(\bigoplus_{k=1}^{n} V^{\otimes k}),$$

where $(i_1, i_2, \ldots, i_n)\colon \bigotimes_{k=1}^{n} T(V^{\otimes k}) \to T(\bigoplus_{k=1}^{n} V^{\otimes k})$ is the composite

$$\bigotimes_{k=1}^{n} T(V^{\otimes k}) \xrightarrow{\bigotimes_{k=1}^{n} T(i_k)} \bigotimes_{k=1}^{n} T(\bigoplus_{k=1}^{n} V^{\otimes k}) \xrightarrow{\text{mult}} T(\bigoplus_{k=1}^{n} V^{\otimes k})$$

and $i_s\colon V^{\otimes s} \to \bigoplus_{k=1}^{n} V^{\otimes k}$ is the canonical inclusion. Observe that because T(V) is cocommutive, each of the maps in the composition is a coalgebra map. Let $h_n\colon T(\bar{J}_n(V)) \to T(\bigoplus_{k=1}^{n} V^{\otimes k})$ be the map of Hopf algebras induced by f_n. It is routine to check that h_n is an isomorphism such that the diagram in the lemma commutes. The assertion follows. □

Theorem 2.7. *The cofiltration*

$$\mathrm{Hom}_{\mathrm{coalg}}(T(-), T(-)) \to \cdots \to \mathrm{Hom}_{\mathrm{coalg}}(J_n(-), T(-)) \to \cdots \to \mathrm{Hom}_{\mathrm{coalg}}(J_0(-), T(-))$$

is a progroup with the property that there is an isomorphism of groups

$$\mathrm{Ker}(\mathrm{Hom}_{\mathrm{coalg}}(J_n(-), T(-)) \to \mathrm{Hom}_{\mathrm{coalg}}(J_{n-1}(-), T(-))) \cong \prod_{q=1}^{\infty} \bigoplus_{l(I)=q} \mathrm{Lie}^I(n).$$

Proof. By Proposition 2.4 and Lemma 2.5, it suffices to show that the homomorphism

$$\mathrm{Hom}_{\mathrm{coalg}}(J_n(-), T(-)) \to \mathrm{Hom}_{\mathrm{coalg}}(J_{n-1}(-), T(-))$$

is an epimorphism. By Lemma 2.6, $T(\bar{J}_{n-1}(V))$ is a natural coalgebra retract of $T(\bar{J}_n(V))$. The assertion follows. □

A natural coalgebra transformation ϕ of graded tensor algebras may not have a geometric realization in general. That is, given a suspension X, there may not exist a map $f\colon \Omega\Sigma X \to \Omega\Sigma X$ such that

$$f_* = \phi\colon H_*(\Omega\Sigma X) = T(\bar{H}_*(X)) \to H_*(\Omega\Sigma X) = T(\bar{H}_*(X)).$$

To be realizable by a map up to a certain dimension often reflects certain properties of X.

Example 2.8. Let $\mathbf{k} = \mathbb{Z}/2$ and let $f_V^n\colon V \to V$ be the $\mathbf{k}$-linear map defined by

$$f_V^n(x) = \begin{cases} x & \text{if } |x| = n \\ 0 & \text{if } |x| \neq n \end{cases}$$

and let $T(f_V^n)\colon T(V) \to T(V)$ be the map of Hopf algebras by the functor T on f_V^n. Let $n \geq 2$ and let $g = T(f_V^n) * T(f_V^{n+1})\colon T(V) \to T(V)$ be the convolution product. Let $X = P^{n+1}(2) = \Sigma^{n-1}\mathbb{R}P^2$ and let u, v be non-zero elements in $H_n(X)$ and H_{n+1}, respectively. Consider the map $g_{\bar{H}_*(X)}\colon H_*(\Omega\Sigma X) = T(u,v) \to H_*(\Omega\Sigma X) = T(u,v)$. Up to dimension $2n+1$, we have

$$g_{\bar{H}_*(X)}(u) = u,\ g_{\bar{H}_*(X)}(v) = v,\ g_{\bar{H}_*(X)}(u^2) = u^2,$$

$$g_{\bar{H}_*(X)}(uv) = g_{\bar{H}_*(X)}(vu) = uv, \ \ g_{\bar{H}_*(X)}([u,v]) = 0.$$

Using this, one can check that there exists a self map ϕ of the $(2n+1)$-skeleton of $\Omega\Sigma X$ such that $\phi_* = g_{\bar{H}_*(X)}$ up to dimension $2n+1$ if and only if the element $[u,v]$ is spherical. Notice that $[u,v]$ is spherical in $H_*(\Omega\Sigma X) = H_*(\Omega P^{n+2}(2))$ if and only if the tangent bundle of S^{n+2} does not have a nowhere zero vector field if and only if n is even. (See [9].)

Now we consider the ungraded case. Although Example 2.8 shows us that an arbitrary natural transformation of graded tensor algebras need not be geometrically realizable, we will see in the next section that graded natural transformations which come from ungraded ones (in the sense of Definition 3.1 are realizable. Let V^u be an ungraded $\mathbf{k}$-module and let $\mathrm{Hom}^u_{\mathrm{coalg}}(T(-),T(-))$ be the set of all of natural morphisms of coalgebras from ungraded tensor algebra $T(V^u)$ to itself. Let $\bar{V}$ be an ungraded $\mathbf{k}$-module with basis $\{x_1,\dots,x_n\}$ Recall that $\mathrm{Lie}(n)$ is the sub $\mathbf{k}$-module of $\bar{V}^{\otimes n}$ generated by the n-fold (ungraded) commutators

$$[[x_{\sigma(1)},x_{\sigma(2)}],\cdots,x_{\sigma(n)}]$$

for $\sigma\in S_n$. Notice that $\bar{V}$ can be regarded as a graded $\mathbf{k}$-module by assigning dimension 2 to its elements. We have the following theorem for the ungraded case.

Corollary 2.9. *The cofiltration*

$$\mathrm{Hom}^u_{\mathrm{coalg}}(T(-),T(-))\to\cdots\to\mathrm{Hom}^u_{\mathrm{coalg}}(J_n(-),T(-))\to\cdots\to\mathrm{Hom}^u_{\mathrm{coalg}}(J_0(-),T(-))$$

is a progroup with the property that there is an isomorphism of groups

$$\mathrm{Ker}(\mathrm{Hom}^u_{\mathrm{coalg}}(J_n(-),T(-))\to\mathrm{Hom}^u_{\mathrm{coalg}}(J_{n-1}(-),T(-)))\xrightarrow{\cong}\mathrm{Lie}(n).$$

3. Geometric Realizations and the Proof of Theorem 1.3

In this section, we give geometric realizations of elements in $\mathrm{Hom}^u_{\mathrm{coalg}}(J_n(-),T(-))$ for $1\le n\le\infty$. Theorem 1.3 will follow from these geometric realizations.

Let V^u be an ungraded $\mathbf{k}$-module and let $\phi_{V^u}\colon T(V^u)\to T(V^u)$ be a natural $\mathbf{k}$-linear map. By Lemma 2.1. there exist $k^n_\sigma\in\mathbf{k}$ for $n\ge1$ and $\sigma\in S_n$ such that

$$\phi_{V^u}(a_1\otimes\cdots\otimes a_n)=\sum_{\sigma\in S_n}k^n_\sigma a_{\sigma(1)}\otimes\cdots\otimes a_{\sigma(n)}$$

for $a_1,\dots,a_n\in V^u$. The set $(k^n_\sigma)_{n\ge1,\sigma\in S_n}$ will be called the coefficient set of ϕ_{V^u}. Let V be a connected graded $\mathbf{k}$-module. Recall that the graded S_n-action on $V^{\otimes n}$ is determined by

$$\tau_{ij}\cdot(\cdots\otimes a_i\otimes\cdots\otimes a_j\otimes\cdots)=(-1)^{|a_i||a_j|}(\cdots\otimes a_j\otimes\cdots\otimes a_i\otimes\cdots)$$

for the generators $\tau_{ij}=(i\ j)\in S_n$.

Definition 3.1. *Let $\phi_{V^u}\colon T(V^u) \to T(V^u)$ be a natural* **k***-linear map of ungraded tensor algebras with the coefficient set $(k^n_\sigma)_{n\geq 1,\sigma\in S_n}$ and let V be a connected graded* **k***-module. The graded* **k***-linear map $\phi^{\mathrm{grade}}_V\colon T(V)\to T(V)$ is defined by*

$$\phi^{\mathrm{grade}}_V(a_1\otimes\cdots\otimes a_n)=\sum_{\sigma\in S_n} k^n_\sigma\sigma\cdot(a_1\otimes\cdots\otimes a_n)$$

for $n\geq 1$ and $a_1\otimes\cdots\otimes a_n\in V^{\otimes n}$.

Lemma 3.2. *Let ϕ_{V^u} and $\phi'_{V^u}\colon T(V^u)\to T(V^u)$ be natural* **k***-linear maps of ungraded tensor algebras. Then*

1) *$\phi^{\mathrm{grade}}_V\colon T(V)\to T(V)$ is a natural* **k***-linear map of graded tensor algebras;*
2) *If ϕ_{V^u} is a natural morphism of coalgebras, then ϕ^{grade}_V is a natural morphism of graded coalgebras;*
3) *The convolution products are preserved. That is,*
$$(\phi*\phi')^{\mathrm{grade}}_V=\phi^{\mathrm{grade}}_V*\phi'^{\ \mathrm{grade}}_V;$$
4) *ϕ^{grade}_V is trivial if and only if ϕ_{V^u} is trivial.*

The proof is immediate.

Corollary 3.3. *The function $\phi_{V^u}\to\phi^{\mathrm{grade}}_V$ is a monomorphism of groups from $\mathrm{Hom}^u_{\mathrm{coalg}}(T(-),T(-))$ to $\mathrm{Hom}_{\mathrm{coalg}}(T(-),T(-))$ and so $\mathrm{Hom}^u_{\mathrm{coalg}}(T(-),T(-))$ is identified with a subgroup of $\mathrm{Hom}_{\mathrm{coalg}}(T(-),T(-))$.*

We need to recall some terminology from [3, 4] before proceeding to prove Theorem 1.3. Let X be a suspension. The group $K_n(X)$ is the subgroup of $[X^n,J(X)]$ generated by the homotopy classes that are represented by the composites

$$X^n\xrightarrow{\ p_i\ }X\hookrightarrow J(X)$$

for $1\leq i\leq n$, where $J(X)$ is the James-construction of X and p_i is the i-th coordinate projection. Let $q_n\colon X^n\to J_n(X)$ be the quotient map. Notice that

$$q_n^*\colon [J_n(X),J(X)]\to[X^n,J(X)]$$

is a monomorphism of groups. Let $H(n)(X)$ denote the intersection $H(n)(X)=[J_n(X),J(X)]\cap K_n(X)$. Then one gets the progroup

$$H(\infty)(X)\twoheadrightarrow\cdots\twoheadrightarrow H(n)(X)\twoheadrightarrow\cdots\twoheadrightarrow H(0)(X).$$

As in [3, 4], we define the Cohen group $K_n(x_1,x_2,\ldots,x_n)$ to be the group generated by $\{x_1,x_2,\ldots,x_n\}$ with relations

(1).

$$[\cdots[[x_{i_1}, x_{i_2}], x_{i_3}], \cdots, x_{i_r}] = 1$$

if $i_s = i_t$ for some $1 \leq s, t \leq r$, where $[a.b] = a^{-1}b^{-1}ab$.

(2).

$$[\cdots[[x_{i_1}^{m_1}, x_{i_2}^{m_2}], x_{i_3}^{m_3}], \cdots, x_{i_r}^{m_r}] = [\cdots[[x_{i_1}, x_{i_2}], x_{i_3}], \cdots, x_{i_r}]^{m_1m_2m_3\cdots m_r}$$

Following Cohen ([3, 4]) we also Let $K_n(x_1, x_2, \ldots, x_n)$ be the Cohen group (see [3, 4]) let $H(n)$ be the equalizer of the projections $p_j\colon K_n(x_1, \ldots, x_n) \to K_{n-1}(x_1, \ldots, x_{n-1})$ for $1 \leq j \leq n$, where the homomorphism $p_j\colon K_n(x_1, \ldots, x_n) \to K_{n-1}(x_1, \ldots, x_{n-1})$ is given by

$$p_j(x_i) = \begin{cases} x_i & \text{for} \quad i < j \\ 1 & \text{for} \quad i = j \\ x_{i-1} & \text{for} \quad i > j. \end{cases}$$

Let $H^R(\infty) = \lim_n H^R(n)$. Then $H(n)$ is the universal group of $H(n)(X)$ in the following sense:

(1). There is a homomorphism

$$e_X\colon H(n) \to H(n)(X) \subseteq [J_n(X), J(X)]$$

for any suspension X.

(2). There exists a suspension X such that

$$e_X\colon H(n) \to [J_n(X), J(X)]$$

is a monomorphism.

(3). Let $\alpha \in H(n)$ and let $f_X\colon J_n(X) \to J(X)$ be a representation the homotopy class $e_X(\alpha)$. Then f_X is functorial up to homotopy.

The progroup $H(\infty)$ is the tower

$$H(\infty) \twoheadrightarrow \cdots \twoheadrightarrow H(n) \twoheadrightarrow \cdots \twoheadrightarrow H(0)$$

with $e_X\colon H(\infty) \to H(\infty)(X) \subseteq [J(X), J(X)]$. Let $\Lambda(n)$ denote the kernel of $H(n) \to H(n-1)$. Let $\bar{V}$ be the free $\mathbb{Z}$-module with basis $\{x_1, \ldots, x_n\}$. Recall that $\mathrm{Lie}^{\mathbb{Z}}(n)$ is the sub $\mathbb{Z}$-module of $\bar{V}^{\otimes n}$ generated by the n-fold commutators $[[x_{\sigma(1)}, x_{\sigma(2)}], \cdots, x_{\sigma(n)}]$ for $\sigma \in S_n$. Recall that, for each $\alpha \in \mathrm{Lie}^{\mathbb{Z}}(n)$, there are unique integers n_τ ($\tau \in S_{n-1}$) such that

$$\alpha = \sum_{\tau \in S_{n-1}} n_\tau[[x_1, x_{\tau(2)}], \cdots, x_{\tau(n)}],$$

where S_{n-1} acts on $\{2, \ldots, n\}$. See [4, 16].

Theorem 3.4 (Cohen [4]). *The group $\Lambda(n)$ is isomorphic to* $\mathrm{Lie}^{\mathbb{Z}}(n)$.

Corollary 3.5 (Theorem 1.3). *Let $f_{V^u}\colon T(V^u)\to T(V^u)$ be a morphism of ungraded coalgebras over $\mathbb{Z}/p\mathbb{Z}$ such that f_{V^u} is a natural transformation of the functor T. Then there is a functorial morphism of graded coalgebras $f^{\text{grade}}\colon T(V)\to T(V)$ for any connected graded module V such that*

(1). *$f^{\text{grade}}_{V^u}=f_{V^u}$ for any ungraded module V^u;*
(2). *$f^{\text{grade}}_{V}\colon T(V)\to T(V)$ is a functorial morphism of bigraded coalgebras, where the bi-grading in $T(V)$ is given in the canonical way;*
(3). *for any suspension X, then there exists a map $\phi_X\colon \Omega\Sigma X\to\Omega\Sigma X$, functorial up to homotopy in X, such that*
$$\phi_{X*}=f^{\text{grade}}_{\bar H_*(X)}\colon H_*(\Omega\Sigma X)=T(\bar H_*(X))\to T(\bar H_*(X)).$$

Proof. By taking homology, the map

$$H_*\colon [J_n(X),J(X)]\longrightarrow \mathrm{Hom}_{\mathrm{coalg}}(H_*(J_n(X)),H_*(J(X)))=$$
$$\mathrm{Hom}_{\mathrm{coalg}}(J_n(\bar H_*(X)),T(\bar H_*(X)))$$

induces a homomorphism

$$\theta\colon H(n)\to \mathrm{Hom}_{\mathrm{coalg}}(J_n(-),T(-))$$

for $0\leq n\leq\infty$ such that the diagram

$$\begin{array}{ccc}
[J_n(X),J(X)] & \xrightarrow{H_*} & \mathrm{Hom}_{\mathrm{coalg}}(J_n(\bar H_*(X)),T(\bar H_*(X)))\\
\uparrow{\scriptstyle e_X} & & \uparrow{\scriptstyle \text{evaluation}}\\
H(n) & \xrightarrow{\theta} & \mathrm{Hom}_{\mathrm{coalg}}(J_n(-),T(-))
\end{array}$$

commutes for any suspension X.

Let τ belong to S_{n-1}, acting on $\{2,\dots,n\}$, and let $\alpha=[[x_1,x_{\tau(2)}],\cdots,x_{\tau(n)}]\in\Lambda(n)$. By the definition of $K_n(X)$, the homotopy class

$$e_X(\alpha)\in H(n)(X)\subseteq[J_n(X),J(X)]\overset{q_n^*}{\hookrightarrow}[X^n,J(X)]$$

is represented by the composite

$$X^n\xrightarrow{q_n}J_n(X)\xrightarrow{\text{pinch}}X^{(n)}\xrightarrow{1\wedge\tau}X^{(n)}\xrightarrow{W_n}J(X),$$

where W_n is the iterated Samelson product taken from left to right and

$$(1\wedge\tau)(x_1\wedge\cdots\wedge x_n)=x_1\wedge x_{\tau(2)}\wedge\cdots\wedge x_{\tau(n)}$$

for $x_1, \dots, x_n \in X$. Let $f_X(\alpha)$ be the composite

$$J_n(X) \xrightarrow{\text{pinch}} X^{(n)} \xrightarrow{1\wedge\tau} X^{(n)} \xrightarrow{W_n} J(X).$$

Consider $f_X(\alpha)_*\colon J_n(\bar{H}_*(X)) \to T(\bar{H}_*(X))$. Then

(1). $f_X(\alpha)_*|_{J_{n-1}(\bar{H}_*(X))}\colon J_{n-1}(\bar{H}_*(X)) \to T(\bar{H}_*(X))$ is trivial;

(2). $f_X(\alpha)_*(a_1\otimes\cdots\otimes a_n) = \beta_n(a_1\otimes\tau\cdot(a_2\otimes\cdots\otimes a_n))$ for $a_1,\dots,a_n \in \bar{H}_*(X)$, where $\beta_n(a_1\otimes\cdots\otimes a_n) = [[a_1,a_2],\cdots,a_n]$, the **graded** n-fold commutator, and S_{n-1} acts on $\bar{H}_*(X)^{\otimes n-1}$ by permuting position with sign.

Thus $\theta([[x_1,x_{\tau(2)}],\cdots,x_{\tau(n)}]) \in \mathrm{Hom}^u_{\mathrm{coalg}}(J_n(-),T(-))$ for any $\tau\in S_{n-1}$ and so

$$\theta(\Lambda(n)) \subseteq \mathrm{Hom}^u_{\mathrm{coalg}}(J_n(-),T(-))$$

for each $1\le n\le\infty$. From the commutative diagram

$$\begin{array}{ccc} H(n) & \xrightarrow{\theta} & \mathrm{Hom}_{\mathrm{coalg}}(J_n(-),T(-)) \\ \downarrow & & \downarrow \\ H(n-1) & \xrightarrow{\theta} & \mathrm{Hom}_{\mathrm{coalg}}(J_{n-1}(-),T(-)), \end{array}$$

by induction, we have

$$\theta(H(n)) \subseteq \mathrm{Hom}^u_{\mathrm{coalg}}(J_n(-),T(-))$$

for $0\le n\le\infty$. By Corollary 2.9, the kernel Γ_n of $\mathrm{Hom}^u_{\mathrm{coalg}}(J_n(-),T(-)) \to \mathrm{Hom}^u_{\mathrm{coalg}}(J_{n-1}(-),T(-))$ is isomorphic to $\mathrm{Lie}(n)$ over $\mathbb{Z}/p$. We have that the map $\theta\colon\Lambda(n)\to\Gamma_n$ is onto. By induction, we have

$$\theta\colon H(n) \to \mathrm{Hom}^u_{\mathrm{coalg}}(J_n(-),T(-))$$

is onto for $0\le n\le\infty$ and so

$$\theta\colon H(\infty) \to \mathrm{Hom}^u_{\mathrm{coalg}}(T(-),T(-))$$

is onto. The assertion follows. □

4. Existence of Minimal Natural Coalgebra Retracts of Tensor Algebras

In this section, we give a proof of Theorem 1.4. The ground ring is a field $\mathbf{k}$ and V means an ungraded $\mathbf{k}$-module. Let $M(V)$ be a functorial submodule of $T(V)$. That is, M is a subfunctor of T from $\mathbf{k}$-modules to $\mathbf{k}$-modules.

Definition 4.1. *A coalgebra $B(V)$ together with natural transformations s_B, r_B, j_B is called a* **natural coalgebra retract of** $T(V)$ **over** $M(V)$ *if:*

(1). *B is a functor from* **k**-*modules to* **k**-*coalgebras;*

(2). *s_B and r_B are natural transformations of coalgebras $s_B\colon B(V) \to T(V)$ and $r_B\colon T(V) \to B(V)$ such that $r_B \circ s_B\colon B(V) \to B(V)$ is the identity map of $B(V)$;*

(3). *j_B is a natural transformation of* **k**-*modules $j_B\colon M(V) \to B(V)$ such that the diagram*

$$\begin{array}{ccccc} B(V) & \xrightarrow{s_B} & T(V) & \xrightarrow{r_B} & B(V) \\ \uparrow{\scriptstyle j_B} & & \uparrow{\scriptstyle j} & & \uparrow{\scriptstyle j_B} \\ M(V) & = & M(V) & = & M(V) \end{array}$$

commutes.

The natural transformation $s_B\colon B(V) \to T(V)$ is called a **coalgebra injection** *and the natural transformation $r_B\colon T(V) \to B(V)$ is called a* **coalgebra retraction**.

Notice that the tensor algebra $T(V)$ is a connected graded Hopf algebra in the canonical way. Let $B(V)$ be a natural coalgebra retract of $T(V)$ over $M(V)$. Let $B_m(V) = B(V) \cap s_B^{-1}(T_m(V))$. From Corollary 2.2 one gets

Lemma 4.2. *Let $B(V)$ be a natural coalgebra retract of $T(V)$ over $M(V)$. Then*

(1). *$B(V) = \oplus_m B_m(V)$ is a connected graded coalgebra;*

(2). *the transformations $s_B\colon B(V) \to T(V)$ and $r_B\colon T(V) \to B(V)$ are natural transformations of graded coalgebras.*

Let $B(V)$ be a natural coalgebra retract of $T(V)$ over $M(V)$. Let $f_B\colon T(V) \to T(V)$ denote the composite

$$T(V) \xrightarrow{r_B} B(V) \xrightarrow{s_B} T(V).$$

Then f_B is idempotent and the diagram

$$\begin{array}{ccc} T(V) & \xrightarrow{f_B} & T(V) \\ \uparrow{\scriptstyle j} & & \uparrow{\scriptstyle j} \\ M(V) & = & M(V) \end{array}$$

commutes. Thus

$$B(V) \cong \operatorname{colim}_{f_B} T(V)$$

as graded coalgebras. Conversely, let $f\colon T(V) \to T(V)$ be a natural transformation of coalgebras such that the diagram

$$\begin{array}{ccc} T(V) & \xrightarrow{f} & T(V) \\ \uparrow{\scriptstyle j} & & \uparrow{\scriptstyle j} \\ M(V) & = & M(V) \end{array}$$

commutes. If the ground field $\mathbf{k}$ is finite, then $\operatorname{colim}_f T(V)$ together with the obvious retraction and inclusions forms a natural coalgebra retract of $T(V)$ over $M(V)$. To draw this conclusion for an arbitrary field $\mathbf{k}$, we need some lemmas.

Lemma 4.3. *Let $\phi_V\colon V^{\otimes k} \to V^{\otimes k}$ be a natural transformation of $\mathbf{k}$-modules and let $\phi_V^n\colon V^{\otimes k} \to V^{\otimes k}$ denote the n-fold self composite of ϕ_V. Then there exists a positive integer $N >> 0$ such that*

1) *N is independent of V;*
2) *the functorial epimorphism*

$$\phi_V\colon \operatorname{Im}(\phi_V^n\colon V^{\otimes k} \to V^{\otimes k}) \twoheadrightarrow \operatorname{Im}(\phi_V^{n+1}\colon V^{\otimes k} \to V^{\otimes k})$$

is a functorial isomorphism for $n \geq N$.

3) *the functorial injection*

$$\operatorname{Im}(\phi_V^{n+1}\colon V^{\otimes k} \to V^{\otimes k}) \hookrightarrow \operatorname{Im}(\phi_V^n\colon V^{\otimes k} \to V^{\otimes k})$$

is a functorial isomorphism for $n \geq N$.

Proof. Let $\bar{V}$ be the $\mathbf{k}$-module with basis $\{x_1, x_2, \dots, x_k\}$. Notice that there is a decreasing sequence

$$\dim_{\mathbf{k}} \bar{V}^{\otimes k} = (\dim_{\mathbf{k}} \bar{V})^k \geq \dim_{\mathbf{k}} \operatorname{Im}(\phi_{\bar{V}}) \geq \dim_{\mathbf{k}} \operatorname{Im}(\phi_{\bar{V}}^2) \geq \cdots .$$

Thus there exists $N >> 0$ such that

$$\dim_{\mathbf{k}} \operatorname{Im}(\phi_{\bar{V}}^n) = \dim_{\mathbf{k}} \operatorname{Im}(\phi_{\bar{V}}^{n+1})$$

for $n \geq N$ and so

$$\operatorname{Im}(\phi_{\bar{V}}^{n+1}\colon \bar{V}^{\otimes k} \to \bar{V}^{\otimes k}) = \operatorname{Im}(\phi_{\bar{V}}^n\colon \bar{V}^{\otimes k} \to \bar{V}^{\otimes k})$$

for $n \geq N$. In particular, there exists $y_n \in \bar{V}^{\otimes k}$ such that

$$\phi_{\bar{V}}^{n+1}(y_n) = \phi_{\bar{V}}^{n}(x_1 \otimes \cdots \otimes x_k)$$

for $n \geq N$. Let V be any $\mathbf{k}$-module and let $a_1, \ldots, a_k$ be k elements in V. Let $f\colon \bar{V} \to V$ be the $\mathbf{k}$-linear map given by $f(x_j) = a_j$ for $1 \leq j \leq k$. From the commutative diagram

$$\begin{array}{ccc} \bar{V}^{\otimes k} & \xrightarrow{f^{\otimes k}} & V^{\otimes k} \\ \big\downarrow \phi_{\bar{V}} & & \big\downarrow \phi_V \\ \bar{V}^{\otimes k} & \xrightarrow{f^{\otimes k}} & V^{\otimes k}, \end{array}$$

we have

$$\phi_V^n(a_1 \otimes \cdots \otimes a_k) = f^{\otimes k}\phi_{\bar{V}}^n(x_1 \otimes x_k) = f^{\otimes k}\phi_{\bar{V}}^{n+1}(y_n) = \phi_V^{n+1}(f^{\otimes k}(y_n))$$

for $n \geq N$. Thus

$$\mathrm{Im}(\phi_V^{n+1}\colon V^{\otimes k} \to V^{\otimes k}) = \mathrm{Im}(\phi_V^n\colon V^{\otimes k} \to V^{\otimes k})$$

for any $\mathbf{k}$-module V; which is assertion 2). Assertion 1) follows from assertion 2). $\square$

Corollary 4.4. *Let $\phi_V\colon V^{\otimes k} \to V^{\otimes k}$ be a natural transformation of $\mathbf{k}$-modules. Then there exists a positive integer $N >> 0$ such that*

1) *N is independent of V;*
2) *the canonical map*

$$\mathrm{Im}(\phi_V^n\colon V^{\otimes k} \to V^{\otimes k}) \longrightarrow \mathrm{colim}_{\phi_V} V^{\otimes k}$$

is a functorial isomorphism for $n \geq N$.

Let $M(V)$ be a functorial $\mathbf{k}$-submodule of $T(V)$. We now generalize the preceding slightly to allow for automorphism of $M(V)$ other than the identity. A natural transformation $f_V\colon T(V) \to T(V)$ is called a **natural transformation over** $M(V)$ if there is a functorial $\mathbf{k}$-linear isomorphism $\bar{f}_V\colon M(V) \to M(V)$ the diagram

$$\begin{array}{ccc} T(V) & \xrightarrow{f_V} & T(V) \\ \big\uparrow j & & \big\uparrow j \\ M(V) & \xrightarrow[\cong]{\bar{f}_V} & M(V) \end{array}$$

commutes.

Theorem 4.5. *Let $M(V)$ be a functorial* $\mathbf{k}$*-submodule of $T(V)$ and let*

$$f_V : T(V) \to T(V)$$

be a natural transformation of coalgebras over $M(V)$. Then $\operatorname{colim}_{f_V} T(V)$ together with the canonical retraction and inclusions forms a natural coalgebra retract of $T(V)$ over $M(V)$.

Proof. By Corollary 4.4, for each $k \geq 1$ there exists an integer $N_k \gg 0$ such that

$$\operatorname{Im}(f_V^n : T_k(V) \to T_k(V)) \longrightarrow \operatorname{colim}_{f_V} T_k(V)$$

is an isomorphism for $n \geq N_k$. We may assume that

$$N_1 < N_2 < N_3 < \cdots .$$

Consider the descending chain of subcoalgebras of $T(V)$

$$T(V) \supseteq \operatorname{Im}(f_V^{N_1} : T(V) \to T(V)) \supseteq \operatorname{Im}(f_V^{N_2} : T(V) \to T(V)) \supseteq \cdots .$$

Let

$$B(V) = \bigcap_{k=1}^{\infty} \operatorname{Im}(f_V^{N_k} : T(V) \to T(V)).$$

Then $B(V)$ is a natural subcoalgebra of $T(V)$ over $M(V)$. Let $\theta\colon B(V) \to \operatorname{colim}_{f_V} T(V)$ be the composite

$$B(V) \hookrightarrow T(V) \longrightarrow \operatorname{colim}_{f_V} T(V).$$

Notice that for each $i \geq 1$,

$$B_i(V) = \bigcap_{k=1}^{\infty} \operatorname{Im}(f_V^{N_k} : T_i(V) \to T_i(V)) = \operatorname{Im}(f_V^{N_i} : T_i(V) \to T_i(V)),$$

by Lemma 4.3. Thus $\theta\colon B(V) \to \operatorname{colim}_{f_V} T(V)$ is an isomorphism of coalgebras. Notice that θ maps the submodule $M(V)$ of $B(V)$ onto the submodule $\operatorname{colim}_{\bar{f}_V} M(V)$ of $\operatorname{colim}_{f_V} T(V)$. The composite $T(V) \to \operatorname{colim}_{f_V} T(V) \xrightarrow{\theta^{-1}} B(V)$ is a functorial retraction over $M(V)$. The assertion follows. □

Let $M(V)$ be a given functorial $\mathbf{k}$-submodule of $T(V)$. Let $\mathcal{R}_M$ denote the class of all natural coalgebra retracts of $T(V)$ over $M(V)$. Let $B(V) \in \mathcal{R}_M$.

Lemma 4.6. *Given $M(V)$, for each $k \geq 1$ there exists a natural coalgebra retract of $T(V)$ over $M(V)$, denoted by $B^{\min_k}(V; M)$, such that the dimension satisfies*

$$\dim_{\mathbf{k}} B_k^{\min_k}(V; M) \leq \dim_{\mathbf{k}} B_k(V)$$

for any $B(V) \in \mathcal{R}_M$ and any $\mathbf{k}$*-module V with $\dim_{\mathbf{k}} V \leq k$.*

Remark 4.7. *We usually write $B^{\min_k}(V)$ instead of $B^{\min_k}(V;M)$ when $M(V)$ is clearly understood.*

Proof. Notice that $B_k(V)$ is a functorial retract of $V^{\otimes k}$ for any $B(V) \in \mathcal{R}_M$ and that for given $B_k(\ )$, $\dim_{\mathbf{k}}(B_k(V))$ only depends on $\dim_{\mathbf{k}}(V)$ if $\dim_{\mathbf{k}}(V)$ is finite. Let $B^{(i)}(V) \subseteq \mathcal{R}_M$ for $1 \le i \le k$ such that

$$\dim_{\mathbf{k}} B_k^{(i)}(V) = \min\{\dim_{\mathbf{k}} B_k(V) \mid B(V) \in \mathcal{R}_M\}$$

if $\dim_{\mathbf{k}} V = i$. Let $s_i \colon B^{(i)}(V) \to T(V)$ be the functorial injection and let $r_i \colon T(V) \to B^{(i)}(V)$ be the functorial retraction for $1 \le i \le k$. Let f_i be the composite

$$T(V) \xrightarrow{r_i} B^{(i)}(V) \xhookrightarrow{s_i} T(V)$$

and let $f_V \colon T(V) \to T(V)$ be the composite

$$T(V) \xrightarrow{f_1} T(V) \xrightarrow{f_2} T(V) \longrightarrow \cdots \xrightarrow{f_k} T(V).$$

Notice that

1) $f_i \circ f_i = f_i$ for $1 \le i \le k$;
2) $\operatorname{Im}(f_i) \cong B^{(i)}(V)$ for $1 \le i \le k$.

Let

$$B^{\min_k}(V) = \operatorname{colim}_{f_V} T(V).$$

Then $B^{\min_k}(V) \in \mathcal{R}_M$ and $B^{\min_k}(V)$ is a natural coalgebra retract of $B^{(i)}(V)$ for each $1 \le i \le k$. The assertion follows. □

Remark 4.8. *$B^{\min_k}(V)$ is natural although not canonical. That is, any particular choice of ordering of the sequence $f_1, \dots, f_k$ gives a natural transformation with the desired properties. Using a different ordering may give a different (although isomorphic) $B^{\min_k}(V)$.*

Lemma 4.9. *Let $f_V \colon T(V) \to T(V)$ be any natural transformation of coalgebras over $M(V)$ and suppose $k \ge 1$. Then the composite*

$$B_k^{\min_k}(V) \xrightarrow{s_{B^{\min_k}}} T_k(V) \xrightarrow{f_V} T_k(V) \xrightarrow{r_{B^{\min_k}}} B_k^{\min_k}(V)$$

is a functorial isomorphism of $\mathbf{k}$-modules for any $B^{\min_k}(V)$ satisfying the conditions of Lemma 4.6.

Proof. Let $g_V \colon T(V) \to T(V)$ denote the composite

$$T(V) \xrightarrow{r_{B^{\min_k}}} B_k^{\min_k}(V) \xhookrightarrow{s_{B^{\min_k}}} T_k(V) \xrightarrow{f_V} T_k(V) \xrightarrow{r_{B^{\min_k}}} B_k^{\min_k}(V) \xhookrightarrow{s^{B^{\min_k}}} T(V).$$

Let

$$A(V) = \operatorname{colim}_{g_V} T(V).$$

Then $A(V)$ is a natural coalgebra retract of $T(V)$ under V with a commutative diagram

$$\begin{array}{ccc} B_k^{\min_k}(V) & \xrightarrow{r_{B^{\min_k}} \circ f_V \circ s_{B^{\min_k}}} & B_k^{\min_k}(V) \\ \downarrow{\phi_1} & & \downarrow{\phi_2} \\ A_k(V) & = & A_k(V), \end{array}$$

where $\phi_1, \phi_2 \colon B^{\min_k}(V) \to A(V)$ are the canonical maps to the colimit.

Notice that $\dim_{\mathbf{k}} B_k^{\min_k}(V)$ is minimum among $\mathbf{k}$-modules V with $\dim_{\mathbf{k}} V \leq k$. Thus both maps

$$\phi_1, \phi_2 \colon B_k^{\min_k}(V) \to A_k(V)$$

are isomorphisms for any $\mathbf{k}$-modules V with $\dim_{\mathbf{k}} V \leq k$ and so the composite

$$B_k^{\min_k}(V) \xrightarrow{s_{B^{\min_k}}} T_k(V) \xrightarrow{f_V} T_k(V) \xrightarrow{r_{B^{\min_k}}} B_k^{\min_k}(V)$$

is a natural isomorphism for any $\mathbf{k}$-modules V with $\dim_{\mathbf{k}} V \leq k$.

Now let V be any finite dimensional $\mathbf{k}$-module. It suffices to show that the map

$$r_{B^{\min_k}} \circ f_V \circ s_{B^{\min_k}} \colon B_k^{\min_k}(V) \to B_k^{\min_k}(V)$$

is an epimorphism. Notice that $B_k^{\min_k}(V)$ is generated, as a graded $\mathbf{k}$-module, by the elements

$$r_{B^{\min_k}}(a_1 \otimes \cdots \otimes a_k),$$

where a_j runs through all elements in V. Let $a_1, \ldots, a_k$ be nonzero elements in V and let $\tilde{V}$ be the $\mathbf{k}$-module with basis $\{x_1, \ldots x_k\}$. Let

$$\phi \colon \tilde{V} \to V$$

be the $\mathbf{k}$-linear map determined by

$$\phi(x_j) = a_j$$

for $1 \leq j \leq k$. Then

$$r_{B^{\min_k}}(a_1 \otimes \cdots \otimes a_k) \in \operatorname{Im}(B^{\min_k}(\phi) \colon B_k^{\min_k}(\tilde{V}) \to B_k^{\min_k}(V)).$$

From the commutative diagram

$$\begin{array}{ccc} B_k^{\min_k}(V) & \xrightarrow{r_{B^{\min_k}} \circ f_V \circ s_{B^{\min_k}}} & B_k^{\min_k}(V) \\ \Big\uparrow B^{\min_k}(\phi) & & \Big\uparrow B^{\min_k}(\phi) \\ B_k^{\min_k}(\tilde{V}) & \xrightarrow[\cong]{r_{B^{\min_k}} \circ f_{\tilde{V}} \circ s_{B^{\min_k}}} & B_k^{\min_k}(\tilde{V}), \end{array}$$

one gets

$$r_{B^{\min_k}}(a_1 \otimes \cdots \otimes a_k) \in \operatorname{Im}(r_{B^{\min_k}} \circ f_V \circ s_{B^{\min_k}} \colon B_k^{\min_k}(V) \to B_k^{\min_k}(V)).$$

Thus the composite

$$B_k^{\min_k}(V) \xrightarrow{s_{B^{\min_k}}} T_k(V) \xrightarrow{f_V} T_k(V) \xrightarrow{r_{B^{\min_k}}} B_k^{\min_k}(V)$$

is a functorial isomorphism for any finite dimensional **k**-module V and so for any **k**-module, which is the assertion. □

Lemma 4.10. *For each $n \geq 1$, there exists a natural coalgebra retract of $T(V)$ over $M(V)$, denoted by $A^{\min_n}(V; M)$, such that for any natural transformation*

$$f_V \colon T(V) \to T(V),$$

of coalgebras over $M(V)$, the composite

$$A_k^{\min_n}(V) \xrightarrow{s_{A^{\min_n}}} T_k(V) \xrightarrow{f_V} T_k(V) \xrightarrow{r_{A^{\min_n}}} A_k^{\min_n}(V)$$

is a functorial isomorphism for $k \leq n$.

Remark 4.11. *As with B we usually write simply $A^{\min_n}(V)$ for $A^{\min_n}(V; M)$.*

Proof. Let $g_k \colon T(V) \to T(V)$ denote the idempotent map

$$T(V) \xrightarrow{r_{B^{min_k}}} B^{\min_k}(V) \xrightarrow{s_{B^{\min_k}}} T(V)$$

for some choice of $B^{\min_k}(V)$. Let $g_V \colon T(V) \to T(V)$ denote the composite

$$T(V) \xrightarrow{g_1} T(V) \xrightarrow{g_2} T(V) \xrightarrow{g_3} \cdots \xrightarrow{g_n} T(V).$$

Let

$$A^{\min_n}(V) = \operatorname{colim}_{g_V} T(V).$$

Then $A^{\min_n}(V)$ is a natural coalgebra retract of $B^{\min_k}(V)$ over $M(V)$ for $1 \leq k \leq n$ and so

$$A_k^{\min_n}(V) \cong B_k^{\min_k}(V)$$

for $1 \leq k \leq n$. The assertion follows from Lemma 4.9 □

Theorem 4.12. *Let $M(V)$ be a functorial* $\mathbf{k}$*-submodule of $T(V)$. Then there exists a natural coalgebra retract of $T(V)$ over $M(V)$, denoted by $A^{\min}(V;M)$, such that $A^{min}(V;M)$ is minimal in the following sense:*

Let $f_V\colon T(V)\to T(V)$ be any natural transformation of coalgebras over $M(V)$. Then the composite

$$A^{\min}(V;M)\xrightarrow{s_{A^{\min}}}T(V)\xrightarrow{f_V}T(V)\xrightarrow{r_{A^{\min}}}A^{\min}(V;M)$$

is a functorial isomorphism of coalgebras.

Proof. For each $n\geq 1$ chose $A^{\min_n}$ satisfying Lemma 4.10. By Lemma 4.10, the composites

$$A_j^{\min_{n+1}}(V)\xrightarrow{s_{A^{\min_{n+1}}}}T_j(V)\xrightarrow{r_{A^{\min_n}}}A_j^{\min_n}(V)\xrightarrow{s_{A^{\min_n}}}T_j(V)\xrightarrow{r_{A^{\min_{n+1}}}}A_j^{\min_{n+1}}(V),$$

$$A_j^{\min_n}(V)\xrightarrow{s_{A^{\min_n}}}T_j(V)\xrightarrow{r_{A^{\min_{n+1}}}}A_j^{\min_{n+1}}(V)\xrightarrow{s_{A^{\min_{n+1}}}}T_j(V)\xrightarrow{r_{A^{\min_n}}}A_j^{\min_n}(V)$$

are isomorphisms for $j\leq n$. Let $f_{n+1}\colon A^{\min_{n+1}}(V)\to A^{\min_n}(V)$ denote the composite

$$A^{\min_{n+1}}(V)\xrightarrow{s_{A^{\min_{n+1}}}}T(V)\xrightarrow{r_{A^{\min_n}}}A^{\min_n}(V).$$

Let $g_{n+1}\colon A^{\min_n}\to A^{\min_{n+1}}$ denote the composite

$$A^{\min_n}\xrightarrow{s_{A^{\min_n}}}T(V)\xrightarrow{r_{A^{\min_{n+1}}}}A^{\min_{n+1}}(V).$$

Then the maps

$$f_{n+1}\colon A_j^{\min_{n+1}}(V)\to A_j^{\min_n}(V),$$
$$g_{n+1}\colon A_j^{\min_n}(V)\to A_j^{\min_{n+1}}(V)$$

are isomorphisms for $j\leq n$.

Let $E(V)$ denote the inverse limit of the sequence

$$\cdots\xrightarrow{f_{n+2}}A^{\min_{n+1}}(V)\xrightarrow{f_{n+1}}A^{\min_n}(V)\xrightarrow{f_n}\cdots\xrightarrow{f_2}A^{\min_1}(V)\xrightarrow{s_{A^{\min_1}}}T(V).$$

Let $\phi_n\colon E(V)\to A^{min_n}(V)$ be the canonical map from the inverse limit to $A^{\min_n}(V)$. Then

$$\phi_n\colon E_j(V)\to A_j^{\min_n}(V)$$

is an isomorphism for $j\leq n$.

Let $F(V)$ denote the direct limit of the sequence

$$T(V)\xrightarrow{r_{A^{\min_1}}}A^{\min_1}(V)\xrightarrow{g_2}\cdots\xrightarrow{g_n}A^{\min_n}(V)\xrightarrow{g_{n+1}}A^{\min_{n+1}}(V)\xrightarrow{g_{n+2}}\cdots.$$

Let $\phi'_n\colon A^{\min_n}(V) \to F(V)$ denote the canonical map from $A^{\min_n}(V)$ to the colimit. Then

$$\phi'_n\colon A_j^{\min_n}(V) \to F_j(V)$$

is an isomorphism for $j \le n$.

Let $\theta_n\colon A^{\min_n}(V) \to A^{\min_n}(V)$ denote the composite

$$A^{\min_n}(V) \xrightarrow{f_n} \cdots \xrightarrow{f_2} A^{\min_1}(V) \xrightarrow{s_{A^{\min_1}}} T(V) \xrightarrow{r_{A^{\min_1}}} A^{\min_1}(V)$$
$$\xrightarrow{g_2} \cdots \xrightarrow{g_n} A^{\min_n}(V).$$

By Lemma 4.10, we have

$$\theta_n\colon A_j^{\min_n}(V) \to A_j^{\min_n}(V)$$

is an isomorphism for $j \le n$. Thus the canonical map

$$E(V) \to F(V)$$

is a natural isomorphism of graded coalgebras over $M(V)$. Let $A^{\min}(V;M) = E(V)$. Then $A^{\min}(V;M)$ is a natural coalgebra retract of $T(V)$ over $M(V)$. Notice that the map $E(V) \to F(V)$ factors through $A^{\min_n}(V)$ for each n. Thus $A^{\min}(V;M)$ is a natural coalgebra retract of $A^{\min_n}(V)$ over $M(V)$ for each n.

By Lemma 4.10, the composite

$$A_j^{\min_n}(V) \xrightarrow{s_{A^{\min_n}}} T_j(V) \xrightarrow{r_{A^{\min}}} A_j^{\min}(V;M) \xrightarrow{s_{A^{\min}}} T_j(V) \xrightarrow{r_{A^{\min_n}}} A_j^{\min_n}(V)$$

is an isomorphism for $j \le n$. Thus $A_j^{\min_n}(V)$ is a functorial retract of $A_j^{\min}(V)$ for $j \le n$ and so, by the previous paragraph, the composites

$$A_j^{\min_n}(V) \xrightarrow{s_{A^{\min_n}}} T_j(V) \xrightarrow{r_{A^{\min}}} A_j^{\min}(V;M),$$

$$A_j^{\min}(V;M) \xrightarrow{s_{A^{\min}}} T_j(V) \xrightarrow{r_{A^{\min_n}}} A_j^{\min_n}(V)$$

are isomorphisms for $j \le n$ if V is a finite dimensional $\mathbf{k}$-module and so they are isomorphisms for $j \le n$ for any $\mathbf{k}$-module V.

Let $f_V\colon T(V) \to T(V)$ be a functorial coalgebra morphism over $M(V)$. Notice that, for each $n \ge 1$, the composite

$$A_j^{\min_n}(V) \xrightarrow{s_{A^{\min_n}}} T_j(V) \xrightarrow{r_{A^{\min}}} A_j^{\min}(V;M) \xrightarrow{s_{A^{\min}}} T_j(V)$$

$$\xrightarrow{f_V} T_j(V) \xrightarrow{r_{A^{\min}}} A_j^{\min}(V) \xrightarrow{s_{A^{\min}}} T_j(V) \xrightarrow{r_{A^{\min_n}}} A_j^{\min_n}(V)$$

is an isomorphism for $j \le n$ by Lemma 4.10. Thus the composite

$$A_j^{\min}(V) \xrightarrow{s_{A^{\min}}} T_j(V) \xrightarrow{f_V} T_j(V) \xrightarrow{r_{A^{\min}}} A_j^{\min}(V)$$

is an isomorphism for any $1 \leq j \leq n$. The assertion follows. □

Corollary 4.13. *Let $M(V)$ be a functorial* **k***-submodule of $T(V)$ and let $B(V)$ be a natural coalgebra retract of $T(V)$ over $M(V)$. Then $A^{\min}(V;M)$ is a natural coalgebra retract of $B(V)$ over $M(V)$.*

Let $M(V)$ be a functorial **k**-submodule of $T(V)$. The coalgebra $A^{\min}(V;M)$ is called the **minimal natural coalgebra retract of $T(V)$ over $M(V)$**. Notice that the minimal natural coalgebra retract of $T(V)$ of $M(V)$ is unique up to isomorphism over $M(V)$ by the universal property in Corollary 4.13. In the special case where $M(V)=V$, let $A^{\min}(V)$ denote the **minimal natural coalgebra retract of $T(V)$ over V**.

5. Some Lemmas on Coalgebras

In this section, the term quasi-Hopf algebra will be used as in [12] to refer to a Hopf algebra in which the multiplication need not be associative, although the comultiplication is assumed to be coassociative.

Let C be a connected graded **k**-coalgebra. Let $P(C)$ denote the set of primitive elements in C. By the proof of Proposition 3.9 in [12], one has the following lemma.

Lemma 5.1. *Let $f\colon A \to B$ be a morphism of connected graded coalgebras over a field. Then $f\colon A_j \to B_j$ is a monomorphism for $j \leq n$ if and only if $P(f)\colon P(A_j) \to P(B_j)$ is a monomorphism for $j \leq n$.*

Let C be a connected graded cocommutative coalgebra and let B be a connected graded cocommutative quasi-Hopf algebra. Notice that the convolution product may not be associative in this setting. There exists a unique left conjugation

$$\chi^l\colon B \to B$$

such that

$$\chi^l * id = \eta_B \circ \epsilon_B.$$

The map $\chi^l\colon B \to B$ is a morphism of graded coalgebras which is given inductively by:

(1). $\chi^l\colon B_0 \to B_0$ is the identity;

(2).

$$\chi^l(x) = -x - \sum \chi(x')x'',$$

when $x \in B_n$ with $n > 0$ and $\psi(x) = x \otimes 1 + 1 \otimes x + \sum x' \otimes x''$.

Similarly, there exists a unique right conjugation $\chi^r\colon B \to B$ such that

$$id * \chi^r = \epsilon_B \circ \eta_B.$$

Lemma 5.2. *Let $p\colon B \to C$ and $s\colon C \to B$ be morphisms of connected graded cocommutative coalgebras such that*

$$p \circ s\colon C \to C$$

is the identity map. Suppose that B is a connected graded cocommutative quasi-Hopf algebra of finite type. Let

$$\phi\colon B \to colim_{(s \circ p \circ \chi^l) * id} B$$

be the canonical map to the colimit. Then the composite

$$B \xrightarrow{\psi} B \otimes B \xrightarrow{\phi \otimes p} colim_{(s \circ p \circ \chi^l) * id} B \otimes C$$

is an isomorphism of graded coalgebras.

Proof. Let f denote the convolution product $(s \circ p \circ \chi^l) * id$ and let $\alpha \in P(B)$. Then

$$f(\alpha) = \alpha - s \circ p(\alpha).$$

Notice that the self map

$$s \circ p\colon B \to B$$

is idempotent. Thus the canonical map

$$P(\mathrm{Im}(f\colon B \to B)) \to P(\mathrm{colim}_f\, B)$$

is a monomorphism and so the canonical coalgebra map

$$\mathrm{Im}(f\colon B \to B) \to \mathrm{colim}_f\, B$$

is a monomorphism. By the proof of Theorem 4.5 the canonical map

$$\phi\colon B \to \mathrm{colim}_f\, B$$

is a split epimorphism of coalgebras. Thus the canonical map

$$\mathrm{Im}(f\colon B \to B) \to \mathrm{colim}_f\, B$$

is an isomorphism. Notice that

$$C \cong \mathrm{Im}(s \circ p\colon B \to B).$$

Let g denote the composite

$$B \xrightarrow{\psi} B \otimes B \xrightarrow{f \otimes (s \circ p)} \mathrm{Im}(f\colon B \to B) \otimes \mathrm{Im}(s \circ p\colon B \to B).$$

Let $\alpha \in P(B)$. Then

$$g(\alpha) = (\alpha - s \circ p(\alpha), s \circ p(\alpha)) \in P(\mathrm{Im}(f\colon B \to B)) \oplus P(\mathrm{Im}(s \circ p\colon B \to B)).$$

Thus

$$g\colon B \to \mathrm{Im}(f\colon B \to B) \otimes \mathrm{Im}(s \circ p\colon B \to B)$$

is a monomorphism and so the composite

$$B \xrightarrow{\psi} B\otimes B \xrightarrow{\phi\otimes p} \operatorname{colim}_f B\otimes C$$

is a monomorphism.

Let h denote the composite

$$\operatorname{Im}(f\colon B\to B)\otimes \operatorname{Im}(s\circ p\colon B\to B) \hookrightarrow B\otimes B \xrightarrow{\mu} B$$
$$\xrightarrow{\psi} B\otimes B \xrightarrow{f\otimes(s\circ p)} \operatorname{Im}(f\colon B\to B)\otimes \operatorname{Im}(s\circ p\colon B\to B).$$

Notice that

$$P(\operatorname{Im}(f\colon B\to B)) = \operatorname{Im}(id - s\circ p\colon P(B)\to P(B)),$$
$$P(\operatorname{Im}(s\circ p\colon B\to B)) = \operatorname{Im}(s\circ p\colon P(B)\to P(B)).$$

Let

$$(\alpha - s\circ p(\alpha), s\circ p(\beta)) \in P(\operatorname{Im}(f\colon B\to B)\otimes \operatorname{Im}(s\circ p\colon B\to B))$$
$$= P(\operatorname{Im}(f\colon B\to B))\oplus P(\operatorname{Im}(s\circ p\colon B\to B)),$$

where $\alpha,\beta\in P(B)$. Then

$$h(\alpha - s\circ p(\alpha), s\circ p(\beta)) = (\alpha - s\circ p(\alpha), s\circ p(\beta)).$$

Thus the coalgebra map h is a monomorphism and so the composite

$$\operatorname{Im}(f\colon B\to B)\otimes \operatorname{Im}(s\circ p\colon B\to B) \hookrightarrow B\otimes B \xrightarrow{\mu} B$$

is a monomorphism. In particular, the Poincaré series

$$\chi(\operatorname{colim}_f B\otimes C) = \chi(\operatorname{Im}(f\colon B\to B)\otimes \operatorname{Im}(s\circ p\colon B\to B)) \le \chi(B).$$

The assertion follows. □

Let B and C be connected graded cocommutative coalgebras and let $p\colon B\to C$ be a morphism of graded coalgebras. B is a C-comodule via the map p and the cotensor product $\mathbf{k}\square_C B$ is a subcoalgebra of B.

Lemma 5.3. *Let $p\colon B\to C$ and let $s\colon C\to B$ be morphisms of connected graded cocommutative coalgebras such that*

$$p\circ s\colon C\to C$$

is the identity map. Let $A = \mathbf{k}\square_C B$ be the cotensor product and let $j\colon A\to B$ be the canonical inclusion. Suppose that B is a connected graded cocommutative quasi-Hopf algebra of finite type. Then the composite

$$A\otimes C \xrightarrow{j\otimes s} B\otimes B \xrightarrow{\mu} B$$

is an isomorphism of graded coalgebras.

Proof. Let f denote the convolution product $(s \circ p \circ \chi^l) * id\colon B \to B$ and let θ denote the composite

$$B \xrightarrow{\psi} B \otimes B \xrightarrow{p \otimes \phi} C \otimes \operatorname{colim}_f B.$$

Notice that there is a commutative diagram

$$\begin{array}{ccc} B & \xrightarrow{\theta} & C \otimes \operatorname{colim}_f B \\ \Big\downarrow p & & \Big\downarrow \mathrm{id}_C \otimes \epsilon_{\operatorname{colim}_f B} \\ C & = & C. \end{array}$$

Thus there is a commutative diagram

$$\begin{array}{ccccc} A = \mathbf{k}\square_C B & \hookrightarrow & B & \xrightarrow{p} & C \\ \Big\downarrow & & \Big\downarrow \theta & & \Big\| \\ \operatorname{colim}_f B & \hookrightarrow & C \otimes \operatorname{colim}_f B & \xrightarrow{\mathrm{id}_C \otimes \epsilon_{\operatorname{colim}_f B}} & C. \end{array}$$

Hence the induced map

$$A \to \operatorname{colim}_f B$$

is a monomorphism. By Lemma 5.2, the map θ is an isomorphism. Thus there is a commutative diagram

$$\begin{array}{ccccc} A & \hookrightarrow & B & \xrightarrow{p} & C \\ \Big\uparrow & & \Big\uparrow \theta^{-1} & & \Big\| \\ \operatorname{colim}_f B & \hookrightarrow & C \otimes \operatorname{colim}_f B & \xrightarrow{\mathrm{id}_C \otimes \epsilon_{\operatorname{colim}_f B}} & C. \end{array}$$

Therefore the induced map

$$\operatorname{colim}_f B \to A$$

is a monomorphism. Notice that both $\operatorname{colim}_f B$ and A are of finite type. Thus

$$A \cong \operatorname{colim}_f B.$$

By examining the restriction to primitives we see that the composite

$$A \otimes C \xrightarrow{j \otimes s} B \otimes B \xrightarrow{\mu} B$$

is a monomorphism. However we have equality

$$\chi(A \otimes C) = \chi(B)$$

of Poincaré series and so the assertion follows. □

6. Functorial Version of the Poincaré-Birkhoff-Witt Theorem

This section contains the proof of Theorem 6.5, which is the main coalgebra decomposition of the paper.

Given an **k**-module V, let $L^{\mathrm{res}}(V)$ denote the free restricted Lie algebra generated by V if the characteristic of **k** is greater than 0 and let it denote $L(V)$, the free Lie algebra generated by V, if **k** is of characteristic 0. Recall that $A^{\min}(V;M)$ denotes the minimal natural coalgebra retract of $T(V)$ over $M(V)$. Thus $A^{\min}(-;M)$ is a functor from **k**-modules to connected graded quasi-Hopf algebras, where $A^{\min}(V;M)$ is graded by assigning dimension 2 to elements of V and the multiplication in $A^{\min}(V;M)$ is given by the composite

$$A^{\min}(V;M)\otimes A^{\min}(V;M) \xhookrightarrow{s_{A^{\min}}\otimes s_{A^{\min}}} T(V)\otimes T(V) \xrightarrow{\mu} T(V) \xrightarrow{r_{A^{\min}}} A^{\min}(V;M).$$

Suppose $n \geq 1$ and let

$$M^n(V) = \oplus_{j=1}^{n} L_j^{\mathrm{res}}(V) \subseteq T(V).$$

Let

$$p\colon T(V) \to A^{\min}(V;M^n)$$

be the **k**-linear map defined by:

(1). $p\colon T_0(V) \to A_0^{\min}(V;M^n)$ is the identity;

(2). $$p(x_1 \cdots x_m) = (\cdots((x_1 \cdot x_2)\cdot x_3)\cdot \cdots \cdot x_m)$$
for $m \geq 1$ and $x_j \in V$, where $\cdot$ is the multiplication in $A^{\min}(V;M^n)$.

Proposition 6.1. *Let $p\colon T(V) \to A^{\min}(V;M^n)$ be defined as above. Then:*

(1). *The map $p\colon T(V) \to A^{\min}(V;M^n)$ is a functorial morphism of coalgebras.*

(2). *The map $p\colon T_j(V) \to A_j^{\min}(V;M^n)$ is an isomorphism for $j \leq n$.*

(3). *The composite*

$$A^{\min}(V;M^n) \xrightarrow{s_{A^{\min}}} T(V) \xrightarrow{p} A^{\min}(V;M^n)$$

is a functorial isomorphism of coalgebras.

(4). *Let I be the kernel of $p\colon T(V) \to A^{\min}(V;M^n)$. Then I is a natural right ideal of $T(V)$.*

(5). *Let* $B^{\max}(V;M^n)=\mathbf{k}\square_{A^{\min}(V;M^n)}T(V)$, *where* p *is used to define the* $A^{\min}(V;M^n)$*-comodule structure on* $T(V)$. *Then* $B^{\max}(V;M^n)$ *is a natural sub Hopf algebra of* $T(V)$.

(6). *There is a functorial isomorphism of coalgebras*

$$\mathbf{k}\otimes_{B^{\max}(V;M^n)}T(V)\cong A^{\min}(V;M^n).$$

Proof. (1) Let $C(V)$ be the connected graded coalgebra defined by:

(i) $C(V)_0=\mathbf{k}$;
(ii) $C(V)_2=V$;
(iii) $P(C(V))=V$.

Then the inclusion map $C(V)\to A^{\min}(V;M^n)$ is a natural map of graded coalgebras. Thus the composite

$$C(V)^{\otimes m}\hookrightarrow (A^{\min}(V;M^n))^{\otimes m}\xrightarrow{\phi} A^{\min}(V;M^n)$$

is a natural morphism of graded coalgebras, where ϕ is m-fold multiplication from left to right. Notice that $T(V)$ is the colimit of the diagram given by

$$C(V)^{\otimes n}\xrightarrow{\iota_j} C(V)^{\otimes(n+1)},$$

for $n\geq 0$ and $1\leq j\leq n+1$, where $\iota_j\colon C(V)^{\otimes n}\to C(V)^{\otimes(n+1)}$ is given by

$$C(V)^{\otimes n}\xrightarrow{id_{C(V)}\otimes\cdots\otimes id_{C(V)}\otimes(\eta_{C(V)}\circ\epsilon_{C(V)})\otimes id_{C(V)}\otimes\cdots\otimes id_{C(V)}} C^{\otimes(n+1)}(V).$$

Assertion (1) follows.

(2) Notice that

$$L_j^{\mathrm{res}}(V)\subseteq A^{\min}(V;M^n)$$

for $1\leq j\leq n$. Thus, by Lemma 5.1,

$$p\colon T_j(V)\to A_j^{\min}(V)$$

is a monomorphism for $j\leq n$. Notice that $A^{\min}(V;M^n)$ is a retract of $T(V)$. Thus for finitely dimensional V

$$p\colon T_j(V)\to A_j^{\min}(V)$$

is an isomorphism for $j\leq n$. Notice that every $\mathbf{k}$-module is a colimit of finitely dimensional $\mathbf{k}$-modules. Assertion (2) follows.

(3) By assertion (1), the map $p\colon T(V)\to A^{\min}(V;M^n)$ is a functorial map of coalgebras over M^n. By Theorem 4.12, the composite

$$A^{\min}(V;M^n)\xrightarrow{s_{A^{\min}}} T(V)\xrightarrow{p} A^{\min}(V;M^n)\xrightarrow{s_{A^{\min}}} T(V)\xrightarrow{r_{A^{\min}}} A^{\min}(V;M^n)$$

is an isomorphism. Assertion (3) follows.

(4) Let $\alpha \in T_k(V)$ be such that

$$p(\alpha) = 0$$

and let $x_j \in V$. Then

$$p(\alpha x_j) = p(\alpha) \cdot x_j = 0.$$

Assertion (4) follows.

(5) Let $\alpha, \beta \in B^{\max}(V; M^n)$ be positive dimensional (homogeneous) elements. Let

$$\psi(\alpha) = \alpha \otimes 1 + 1 \otimes \alpha + \sum \alpha' \otimes \alpha'',$$

$$\psi(\beta) = \beta \otimes 1 + 1 \otimes \beta + \sum \beta' \otimes \beta''.$$

Then, in $T(V)$,

$$\begin{aligned} \psi(\alpha\beta) &= \alpha\beta \otimes 1 + 1 \otimes \alpha\beta + \alpha \otimes \beta \\ &+ (-1)^{|\alpha||\beta|}\beta \otimes \alpha + \sum((-1)^{|\alpha''||\beta|}\alpha'\beta \otimes \alpha'' + \alpha' \otimes \alpha''\beta) \\ &+ \sum(\alpha\beta' \otimes \beta'' + (-1)^{|\alpha||\beta'|}\beta' \otimes \alpha\beta'') + \sum(-1)^{|\alpha''||\beta'|}\alpha'\beta' \otimes \alpha''\beta''. \end{aligned}$$

Notice that because of cocommutativity

$$\alpha, \alpha', \alpha'', \beta, \beta', \beta'' \in I(B) \subseteq I = \ker(p).$$

By assertion (4), I is a right ideal. Thus

$$(p \otimes id_{T(V)}) \circ \psi(\alpha\beta) = 1 \otimes \alpha\beta$$

in $C \otimes T(V)$ and so $\alpha\beta \in B^{\max}(V; M^n)$. Assertion (5) follows.

(6) By assertions (4) and (5) the map

$$p\colon T(V) \to A^{\min}(V; M^n)$$

factors through $\mathbf{k} \otimes_{B^{\max}(V;M^n)} T(V)$. Let

$$\bar{p}\colon \mathbf{k} \otimes_{B^{\max}(V;M^n)} T(V) \to A^{\min}(V; M^n)$$

denote the resulting epimorphism. If V is finitely dimensional, then, by Lemma 5.3, the Poincaré series

$$\chi(A^{\min}(V; M^n)) = \chi(T(V))/\chi(B^{\max}(V; M^n)).$$

By Theorem 4.4 in [12], we have

$$\chi(\mathbf{k} \otimes_{B^{\max}(V;M^n)} T(V)) = \chi(T(V))/\chi(B^{\max}(V; M^n)).$$

Thus $\bar{p}\colon \mathbf{k} \otimes_{B^{\max}(V;M^n)} T(V) \to A^{\min}(V; M^n)$ is an isomorphism if V is finitely dimensional. Notice that every $\mathbf{k}$-module is a colimit of finitely dimensional $\mathbf{k}$-modules. Assertion (6) follows. □

By Lemma 5.3, the sub-Hopf algebra $B^{\max}(V;M^n)$ is a functorial coalgebra retract of $T(V)$ and thus, by Lemma 4.2, it is a functorial graded coalgebra retract of $T(V)$. Notice that

$$B_j^{\max}(V;M^n)=0$$

for $0<j<n+1$. Thus

$$B^{\max}_{n+1}(V;M^n)=P(B^{\max}_{n+1}(V;M^n))\subseteq P(T_{n+1}(V))=L^{\mathrm{res}}_{n+1}(V).$$

Let $L^{\max}_{n+1}(V)$ denote $B^{\max}_{n+1}(V;M^n)$ for $n\geq 1$. Recall that $\mathbf{k}(S_n)$ acts functorially on the right on $V^{\otimes n}$ by permuting positions. For $\lambda\in\mathbf{k}(S_n)$, we also use λ to denote the functorial map $V^{\otimes n}\to V^{\otimes n}$ given by the action of λ.

Define $\beta_n : V^{\otimes n}\to V^{\otimes n}$ by $\beta_n(x_1\otimes\cdots\otimes x_n)=[[x_1,x_2],\ldots,x_n]$.

Lemma 6.2. *Suppose $n\geq 2$ and let $L^{\max}_n(V)$ be defined as above. Then:*

(1). *$L^{\max}_n(V)$ is a functorial retract of $L_n(V)$;*

(2). *there exists an element*

$$\lambda^{\max}_n\in\mathbf{k}(S_n)$$

such that:

(i) *$\beta_n\circ\lambda^{\max}_n\circ\beta_n\circ\lambda^{\max}_n=\beta_n\circ\lambda^{\max}_n$;*

(ii) *there is a functorial isomorphism*

$$L^{\max}_n(V)\cong\operatorname{colim}_{\beta_n\circ\lambda^{\max}_n}V^{\otimes n}=\operatorname{Im}(\beta_n\circ\lambda^{\max}_n : V^{\otimes n}\to V^{\otimes n})$$

for $\mathbf{k}$-modules V.

Proof. Let

$$j\colon B^{\max}(V;M^{n-1})\to T(V)$$

be the canonical functorial injection and let

$$r\colon T(V)\to B^{\max}(V;M^{n-1})$$

be a functorial coalgebra retraction. Let f denote the composite $j\circ r\colon T(V)\to T(V)$. Notice that f is idempotent and that

$$B^{\max}(V;M^{n-1})=\operatorname{colim}_f T(V).$$

By Lemma 2.1, there exists an element $\theta\in\mathbf{k}(S_n)$ such that

$$\theta=f|_{T_n(V)}\colon T_n(V)=V^{\otimes n}\to V^{\otimes n}.$$

Let $\bar V$ be the $\mathbf{k}$-module with basis $\{x_1,\ldots,x_n\}$. Let γ_n denote the submodule of $T_n(\bar V)=\bar V^{\otimes n}$ spanned by the elements

$$x_{\sigma(1)}\cdots x_{\sigma(n)}$$

as σ runs through S_n. Consider the map $\theta\colon\bar V^{\otimes n}\to\bar V^{\otimes n}$. Notice that

$$\operatorname{Im}(\theta\colon\bar V^{\otimes n}\to\bar V^{\otimes n})=\operatorname{Im}(f\colon T_n(\bar V)\to T_n(\bar V))=L^{\max}_n(\bar V)\subseteq L^{\mathrm{res}}_n(\bar V).$$

Thus

$$\theta(x_1 \cdots x_n) \in \gamma_n \cap L_n^{\mathrm{res}}(\bar{V}) = \gamma_n \cap L_n(\bar{V}) = \mathrm{Lie}(n).$$

Thus

$$\theta(x_1 \cdots x_n) = \sum_{\tau \in S_{n-1}} k_\tau [[x_1, x_{\tau(2)}, \cdots, x_{\tau(n)}]$$

for some $k_\tau \in \mathbf{k}$, where S_{n-1} acts on $\{2, 3, \ldots, n\}$. For each $\tau \in S_{n-1}$, let $\bar{\tau} \in S_n$ be defined by

$$\bar{\tau}(1) = 1, \bar{\tau}(j) = \tau(j)$$

for $2 \leq j \leq n$. Now let

$$\lambda_n^{\max} = \sum_{\tau \in S_{n-1}} k_\tau \bar{\tau} \in \mathbf{k}(S_n).$$

Let

$$\theta' = \beta_n \circ \lambda_n^{\max} \colon V^{\otimes n} \longrightarrow V^{\otimes n}.$$

In the case $V = \bar{V}$,

$$\theta(x_1 \cdots x_n) = \theta'(x_1 \cdots x_n).$$

Let $e \colon \mathbf{k}(S_n) \to \mathrm{Hom}_{\mathbf{k}}(\bar{V}^{\otimes n} \to \bar{V}^{\otimes n})$ be the representation map of the $\mathbf{k}(S_n)$-action on $\bar{V}^{\otimes n}$. Then

$$e(\theta) = e(\theta')$$

in $\mathrm{Hom}_{\mathbf{k}}(\bar{V}^{\otimes n}, \bar{V}^{\otimes n})$. Notice that the representation map $e \colon \mathbf{k}(S_n) \to \mathrm{Hom}_{\mathbf{k}}(\bar{V}^{\otimes n}, \bar{V}^{\otimes n})$ is faithful. Thus

$$\theta = \theta' = \beta_n \circ \lambda_n^{\max} \colon V^{\otimes n} \longrightarrow V^{\otimes n}.$$

Notice that

$$\theta \circ \theta = f \circ f = f = \theta.$$

The assertions follow. □

Lemma 6.3. *Suppose $n \geq 2$. Then there is a functorial isomorphism of coalgebras*

$$A^{\min}(V; M^n) \cong A^{\min}(V; L_n^{\max}) \otimes A^{\min}(V; M^{n-1})$$

for any $\mathbf{k}$-module V.

Proof. Consider the commutative diagram of short exact sequences of coalgebras

$$\begin{array}{ccccc}
B^{\max}(V; M^{n-1}) & \overset{j}{\hookrightarrow} & T(V) & \xrightarrow{\quad p \quad} & A^{\min}(V; M^{n-1}) \\
\uparrow & & \uparrow{\scriptstyle s_{A^{\min}(V;M^n)}} & & \| \\
\mathbf{k}\square_{A^{\min}(V;M^{n-1})} A^{\min}(V; M^n) & \overset{j}{\hookrightarrow} & A^{\min}(V; M^n) & \xrightarrow{p \circ s_{A^{\min}(V;M^n)}} & A^{\min}(V; M^{n-1}).
\end{array}$$

Notice that the composite

$$A^{\min}(V;M^{n-1}) \xrightarrow{s_{A^{\min}(V;M^{n-1})}} T(V) \xrightarrow{r_{A^{\min}(V;M^{n})}} A^{\min}(V;M^{n})$$

$$\xrightarrow{p\circ s_{A^{\min}(V;M^{n})}} A^{\min}(V;M^{n-1})$$

is a natural map of coalgebra **over** $M^{n-1}(V)=\oplus_{1\leq j\leq n-1}L_j^{\mathrm{res}}(V)$. By Theorem 4.12, this composite is a functorial isomorphism of graded coalgebras. Thus there is a cross-section map of coalgebras

$$\tilde{s}\colon A^{\min}(V;M^{n-1})\to A^{\min}(V;M^{n})$$

such that the composite

$$p\circ s_{A^{\min}(V;M^{n})}\circ\tilde{s}\colon A^{\min}(V;M^{n-1})\to A^{\min}(V;M^{n-1})$$

is the identity map. Let $D(V)$ denote the cotensor product $\mathbf{k}\square_{A^{\min}(V;M^{n-1})}A^{\min}(V;M^{n})$. By Lemma 5.3, $D(V)$ is a natural coalgebra

retract of $A^{\min}(V;M^{n})$ if V is finite dimensional. Using a colimit argument, we conclude that $D(V)$ is a natural coalgebra retract of $A^{\min}(V;M^{n})$ for any $\mathbf{k}$-module V and so $D(V)$ is a natural coalgebra retract of $T(V)$ for any $\mathbf{k}$-module V. Notice that because of Proposition 6.1(2)

$$D_n(V)=B_n^{\max}(V;M^{n-1})=L_n^{\max}(V).$$

Thus $D(V)$ is a natural coalgebra retract of $T(V)$ over $L_n^{\max}(V)$ and so $A^{\min}(V;L_n^{\max})$ is a natural coalgebra retract of $D(V)$. More precisely, there exist functorial maps of coalgebras $f\colon A^{\min}(V;L_n^{\max})\to D(V)$ and $g\colon D(V)\to A^{\min}(V;L_n^{\max})$ such that

(1). $f|_{A_n^{\min}(V;L_n^{\max})}\colon A_n^{\min}(V;L_n^{\max})=L_n^{\max}(V)\to D_n(V)=L_n^{\max}(V)$ and $g|_{D_n(V)}\colon D_n(V)=L_n^{\max}(V)\to A_n^{\min}(V)=L_n^{\max}(V)$ are the identity maps;

(2). the composite $g\circ f\colon A^{\min}(V;L_n^{\max})\to A^{\min}(V;L_n^{\max})$ is the identity map.

By Lemmas 5.2 and 5.3, there is a functorial coalgebra retraction

$$\tilde{r}\colon A^{\min}(V;M^{n})\to D(V)$$

such that the composite

$$A^{\min}(V;M^{n})\xrightarrow{\psi}A^{\min}(V;M^{n})\otimes A^{\min}(V;M^{n})\xrightarrow{\tilde{r}\otimes(p\circ s_{A^{\min}(V;M^{n})})}$$

$$D(V)\otimes A^{\min}(V;M^{n-1})\xrightarrow{j\otimes\tilde{s}}A^{\min}(V;M^{n})\otimes A^{\min}(V;M^{n})\xrightarrow{\mu}A^{\min}(V;M^{n})$$

is a functorial isomorphism of coalgebras for any finitely dimensional $\mathbf{k}$-module V and so for any $\mathbf{k}$-module V. Let $\phi\colon T(V)\to T(V)$ be the composite

$$T(V)\longrightarrow A^{\min}(V;M^{n})\xrightarrow{\tilde{r}\otimes(p\circ s_{A^{\min}(V;M^{n})})\circ\psi}D(V)\otimes A^{\min}(V;M^{n-1})$$

$$\xrightarrow{g\otimes id} A^{\min}(V;L^{\max})\otimes A^{\min}(V;M^{n-1}) \xrightarrow{f\otimes id} D(V)\otimes A^{\min}(V;M^{n-1})$$
$$\xrightarrow{\mu\circ(j\otimes\tilde{s})} A^{\min}(V;M^n) \longrightarrow T(V).$$

Then $\phi|_{T_j(V)}\colon T_j(V)\to T_j(V)$ is a natural isomorphism for $j\le n$. Thus

$$\operatorname{colim}_\phi T(V)$$

is a natural coalgebra retract of $T(V)$ over M^n and so $A^{\min}(V;M^n)$ is a natural coalgebra retract of $\operatorname{colim}_\phi T(V)$. Notice that the map $\phi\colon T(V)\to T(V)$ factors through $A^{\min}(V;L_n^{\max})\otimes A^{\min}(V;M^{n-1})$. Thus $\operatorname{colim}_\phi T(V)$ is a coalgebra retract of $A^{\min}(V;L_n^{\max})\otimes A^{\min}(V;M^{n-1})$ and so $A^{\min}(V;M^n)$ is natural coalgebra retract of $A^{\min}(V;L_n^{\max})\otimes A^{\min}(V;M^{n-1})$. In particular, the Poincaré series of primitive submodules satisfy

$$\begin{aligned}\chi(P(A^{\min}(V;M^n))) &= \chi(P(D(V)))+\chi(P(A^{\min}(V;M^{n-1})))\\ &\le \chi(P(A^{\min}(V;L_n^{\max})))+\chi(P(A^{\min}(V;M^{n-1})))\end{aligned}$$

for any finitely dimensional $\mathbf{k}$-module V and so

$$\chi(P(D(V)))\le\chi(P(A^{\min}(V;L_n^{\max})))$$

for any finitely dimensional $\mathbf{k}$-module V. Notice that

$$f\colon P(A^{\min}(V;L_n^{\max}))\to P(D(V))$$

is a monomorphism. Thus $f\colon P(A^{\min}(V;L_n^{\max}))\to P(D(V))$ is a functorial isomorphism for any finitely dimensional $\mathbf{k}$-module V. Recall that

$$g\circ f\colon A^{\min}(V;L_n^{\max})\to A^{\min}(V;L_n^{\max})$$

is the identity map and so

$$g\colon P(D(V))\to P(A^{\min}(V;L_n^{\max})$$

is an epimorphism. Thus $g\colon P(D(V))\to P(A^{\min}(V;L_n^{\max}))$ is a natural isomorphism for any finitely dimensional $\mathbf{k}$-module V. Therefore

$$g\colon D(V)\to A^{\min}(V;L_n^{\max})$$

is a functorial monomorphism and thus an isomorphism for any finitely dimensional $\mathbf{k}$-module V. Therefore $g\colon D(V)\to A^{\min}(V;L_n^{\max})$ is a functorial isomorphism of coalgebras for any $\mathbf{k}$-module V. The assertion follows. □

Corollary 6.4. *The functor $A^{\min}(-;L_n^{\max})$ has the property*

$$A_j^{\min}(V;L_n^{\max})=\begin{cases}0 & \text{for } 0<j<n;\\ L_n^{\max}(V) & \text{for } j=n.\end{cases}$$

Let $L_1^{\max}(V)$ denote V. A functorial coalgebra decomposition of tensor algebras is as follows.

Theorem 6.5 (Functorial Poincaré-Birkhoff-Witt). *There exists a functorial isomorphism of coalgebras*

$$T(V) \cong \bigotimes_{n=1}^{\infty} A^{\min}(V; L_n^{\max})$$

for any **k**-*module* V.

Proof. By Lemmas 6.3 and induction, there is a functorial isomorphism of coalgebras

$$\phi_n \colon \bigotimes_{j=1}^{n} A^{\min}(V; L_j^{\max}) \to A^{\min}(V; M^n)$$

such that the diagram

$$\begin{array}{ccc}
\bigotimes_{j=1}^{n} A^{\min}(V; L_j^{\max}) & \xrightarrow[\cong]{\phi_n} & A^{\min}(V; M^n) \\
\uparrow & & \uparrow \tilde{s}_n \\
\bigotimes_{j=1}^{n-1} A^{\min}(V; L_j^{\max}) & \xrightarrow[\cong]{\phi_{n-1}} & A^{\min}(V; M^{n-1})
\end{array}$$

commutes, where $\tilde{s}_n \colon A^{\min}(V; M^{n-1}) \to A^{\min}(V; M^n)$ is an injection of coalgebras. Notice that the colimit of the sequence

$$A^{\min}(V) = A^{\min}(V; M^1) \hookrightarrow A^{\min}(V; M^2) \hookrightarrow A^{\min}(V; M^3) \hookrightarrow \cdots.$$

is isomorphic to $T(V)$. The assertion follows. $\square$

Theorem 6.6. *Suppose $n \geq 2$ and let $\lambda \in \mathbf{k}(S_n)$ be any element. Then the colimit*

$$\operatorname{colim}_{\beta_n \circ \lambda} V^{\otimes n}$$

is a functorial retract of $L_n^{\max}(V)$.

Proof. Let $M(V)$ denote $\operatorname{colim}_{\beta_n \circ \lambda} V^{\otimes n}$. Let $p_m \colon V^{\otimes n} \to \operatorname{colim}_{\beta_n \circ \lambda} V^{\otimes n}$ be the canonical map from the m-th term in the sequence

$$V^{\otimes n} \xrightarrow{\beta_n \circ \lambda} V^{\otimes n} \xrightarrow{\beta_n \circ \lambda} V^{\otimes n} \xrightarrow{\beta_n \circ \lambda} \cdots$$

to the colimit. By Lemma 4.3, the **k**-module $\operatorname{colim}_{\beta_n \circ \lambda} V^{\otimes n}$ is a functorial retract of $V^{\otimes n}$. Thus there is a functorial submodule

$$M'(V) \subseteq L_n(V)$$

such that the composite

$$M'(V) \hookrightarrow L_n(V) \hookrightarrow V^{\otimes n} \xrightarrow{p_2} M(V)$$

is a functorial isomorphism.

Now the functorial injection $j\colon M'(V) \subseteq V^{\otimes n}$ induces a unique map of algebras

$$\phi\colon T(M'(V)) \to T(V)$$

such that $\phi|_{M'(V)}\colon M'(V) \to V^{\otimes n} \hookrightarrow T(V)$ is the functorial injection. Since $M'(V) \subseteq L_n(V) \subseteq PT(V)$ the functorial map $\phi\colon T(M'(V)) \to T(V)$ is a map of Hopf algebras.

Let $H_n\colon T(V) \to T(V^{\otimes n})$ be the James-Hopf map [8] and let

$$\phi'\colon T(V) \to T(M(V))$$

be the composite

$$T(V) \xrightarrow{H_n} T(V^{\otimes n}) \xrightarrow{T(p_2)} T(M(V)).$$

Then the map $\phi'\colon T(V) \to T(M(V))$ is a functorial map of coalgebras. Notice that the diagram

$$\begin{array}{ccc} T(L_n(V)) & \hookrightarrow & T(V) \\ \downarrow & & \downarrow H_n \\ T(V^{\otimes n}) & = & T(V^{\otimes n}) \end{array}$$

commutes by [8, Proposition 5.1] or [17, Corollary1.2]. Thus there is a commutative diagram

$$\begin{array}{ccc} T(L_n(V)) & \hookrightarrow & T(V) \\ \uparrow & & \downarrow \phi' \\ T(M'(V)) & \xrightarrow{T(p_2)\circ j} & T(M(V)). \end{array}$$

Notice that $p_2 \circ j\colon M'(V) \to M(V)$ is an isomorphism. Thus $T(M(V))$ is a natural coalgebra retract of $T(V)$. By Lemma 5.2, there exists a natural coalgebra retract $A(V)$ of $T(V)$ such that there is a functorial isomorphism of graded coalgebras

$$T(V) \cong T(M(V)) \otimes A(V)$$

for any finite dimensional $\mathbf{k}$-module V and so for any $\mathbf{k}$-module V. Notice that

$$T_j(V) \cong A_j(V)$$

for $j \leq n-1$. Thus $A(V)$ is a natural coalgebra retract of $T(V)$ over $M^{n-1}(V) = \oplus_{1\leq j\leq n-1} L_j^{\mathrm{res}}(V)$. By Corollary 4.13, $A^{\min}(V; M^{n-1})$ is a functorial coalgebra retract of $A(V)$ over $M^{n-1}(V)$. By Lemma 5.2, there exists a functorial coalgebra $D(V)$ such that there is a functorial isomorphism of coalgebras

$$A(V) \cong D(V) \otimes A^{\min}(V; M^{n-1})$$

for any finite dimensional $\mathbf{k}$-module V and so for any $\mathbf{k}$-module V. Thus there is a functorial isomorphism of coalgebras

$$T(V) \cong T(M(V)) \otimes D(V) \otimes A^{\min}(V; M^{n-1})$$

and so there is a functorial isomorphism of coalgebras

$$\theta\colon B^{\max}(V; M^{n-1}) \otimes A^{\min}(V; M^{n-1}) \xrightarrow{\cong} T(M(V)) \otimes D(V) \otimes A^{\min}(V; M^{n-1}).$$

Let

$$q = \epsilon_{T(M(V))\otimes D(V)} \otimes id_{A^{\min}(V;M^{n-1})} \quad \colon T(M(V)) \otimes D(V) \otimes A^{\min}(V; M^{n-1}) \to A^{\min}(V; M^{n-1})$$

be the canonical projection and let $\bar{B}(V)$ denote the cotensor product

$$\bar{B}(V) = \mathbf{k}\square_{A^{\min}(V;M^{n-1})} \left(B^{\max}(V; M^{n-1}) \otimes A^{\min}(V; M^{n-1})\right),$$

where the $A^{\min}(V; M^{n-1})$-comodule structure on $B^{\max}(V; M^{n-1}) \otimes A^{\min}(V; M^{n-1})$ is induced by the composite of the coalgebra maps

$$q \circ \theta\colon B^{\max}(V; M^{n-1}) \otimes A^{\min}(V; M^{n-1}) \longrightarrow A^{\min}(V; M^{n-1}).$$

Let

$$i = \eta_{B^{\max}(V;M^{n-1})} \otimes id_{A^{\min}(V;M^{n-1})}\colon A^{\min}(V; M^{n-1}) \to B^{\max}(V; M^{n-1}) \otimes A^{\min}(V; M^{n-1})$$

be the canonical inclusion map and let $\theta'\colon A^{\min}(V; M^{n-1}) \to A^{\min}(V; M^{n-1})$ denote the composite

$$\theta' = q \circ \theta \circ i\colon A^{\min}(V; M^{n-1}) \to A^{\min}(V; M^{n-1}).$$

Then

$$\theta'\colon A_j^{\min}(V; M^{n-1}) \to A_j^{\min}(V; M^{n-1})$$

is an isomorphism for $j \leq n-1$. Thus, by Theorem 4.12, the functorial coalgebra map $\theta'\colon A^{\min}(V; M^{n-1}) \to A^{\min}(V; M^{n-1})$ is an isomorphism. Thus the short exact sequence of coalgebras

$$\bar{B}(V) \hookrightarrow B^{\max}(V; M^{n-1}) \otimes A^{\min}(V; M^{n-1}) \xrightarrow{q \circ \theta} A^{\min}(V; M^{n-1})$$

splits functorially by Lemma 5.3 and so

$$B^{\max}(V; M^{n-1}) \otimes A^{\min}(V; M^{n-1}) \cong \bar{B}(V) \otimes A^{\min}(V; M^{n-1})$$

as coalgebras. By the commutative diagram of short exact sequences of coalgebras

$$\begin{array}{ccccc} T(M(V))\otimes D(V) & \hookrightarrow & T(M(V))\otimes D(V)\otimes A^{\min}(V;M^{n-1}) & \xrightarrow{q} & A^{\min}(V;M^{n-1}) \\ \uparrow & & \uparrow{\scriptstyle\theta} & & \| \\ \bar{B}(V) & \xhookrightarrow{i'} & B^{\max}(V;M^{n-1})\otimes A^{\min}(V;M^{n-1}) & \xrightarrow{q\circ\theta} & A^{\min}(V;M^{n-1}), \end{array}$$

the resulting functorial coalgebra map $\bar{B}(V)\to T(M(V))\otimes D(V)$ is a monomorphism. Notice that both $\bar{B}(V)$ and $T(M(V))\otimes D(V)$ have the same Poincaré series if V is finite dimensional. Thus there is a functorial isomorphism of coalgebras

$$\bar{B}(V)\cong T(M(V))\otimes D(V)$$

for any finite dimensional $\mathbf{k}$-module V and so for any $\mathbf{k}$-module V. The assertion follows. □

Corollary 6.7. *Suppose the ground field* $\mathbf{k}$ *is of characteristic* p *and let* $n\geq 1$ *be an integer such that* n *is not divisible by* p. *Then*

$$L_n^{\max}(V)=L_n(V).$$

If $\mathrm{char}(\mathbf{k})=0$, *then* $L_n^{\max}(V)=L_n(V)$ *for all* $n\geq 1$.

Proof. Notice that $L_n^{max}(V)\subseteq L_n(V)$ and $L_n(V)\cong\operatorname{colim}_{\beta_n}V^{\otimes n}$ by the Dynkin-Specht-Weber formula

$$\beta_n\circ\beta_n=n\beta_n.$$

The assertion follows. □

Further functorial coalgebra decomposition of $A^{\min}(V;L_n^{\max})$ can be given by using the following proposition.

Proposition 6.8. *Let* M *be a subfunctor of* L_n *from* $\mathbf{k}$*-modules to* $\mathbf{k}$*-modules. Suppose that*

1) M *is a retract of the functor* $L_n^{\max}$;
2) *there are functors* M' *and* M'' *from* $\mathbf{k}$*-modules to* $\mathbf{k}$*-modules such that there is a functorial isomorphism of* $\mathbf{k}$*-modules*

$$M(V)\cong M'(V)\oplus M''(V).$$

Then there is a functorial isomorphism of coalgebras

$$A^{\min}(V;M)\cong A^{\min}(V;M')\otimes A^{\min}(V;M'').$$

Proof. One may assume that M' is a subfunctor of M from $\mathbf{k}$-modules to $\mathbf{k}$-modules. Notice that

$$M'(V) \subseteq M(V) \subseteq A^{\min}(V; M)$$

and $A^{\min}(V; M)$ is a natural coalgebra retract of $T(V)$. Thus $A^{\min}(V; M')$ is a natural coalgebra retract of $A^{\min}(V; M)$ by Corollary 4.13 and so there is a functorial coalgebra decomposition

$$A^{\min}(V; M) \cong A^{\min}(V; M') \otimes B(V)$$

for any finite dimensional $\mathbf{k}$-module V by Lemma 5.3, where

$$B(V) = \mathbf{k}\square_{A^{\min}(V;M')} A^{\min}(V; M)$$

and $A^{\min}(V; M)$ is an $A^{\min}(V; M')$-comodule via a given (functorial) coalgebra retraction from $A^{\min}(V; M)$ to $A^{\min}(V; M')$. By using colimit arguments, this decomposition holds for any $\mathbf{k}$-module V.

Let $r_{L_n^{\max}} \colon V^{\otimes n} \to L_n^{\max}(V)$ and $r_M \colon L_n^{\max}(V) \to M(V)$ be functorial retractions. Notice that the composite of functorial coalgebra maps

$$T(M(V)) \hookrightarrow T(L_n^{\max}(V)) \hookrightarrow T(L_n(V)) \hookrightarrow T(V)$$

$$\xrightarrow{H_n} T(V^{\otimes n}) \xrightarrow{T(r_{L_n^{\max}})} T(L_n^{\max}(V)) \xrightarrow{T(r_M)} T(M(V)).$$

is multiplicative, where H_n is the James-Hopf map. Thus this composite is determined by the restriction to $M(V)$ and so it is an isomorphism. Thus $T(M(V))$ is a functorial coalgebra retract of $T(V)$ over $M(V)$ and so $A^{\min}(V; M)$ is a functorial coalgebra retract of $T(M(V))$. Thus $A_j^{\min}(V; M) = 0$ for $0 < j < n$ and $A_n^{\min}(V; M) = M(V)$. Similarly, $A_j^{\min}(V; M') = 0$ for $0 < j < n$ and $A_n^{\min}(V; M') = M'(V)$. Thus $B_j(V) = 0$ for $0 < j < n$ and $B_n(V) \cong M''(V)$. By Corollary 4.13, $A^{\min}(V; M'')$ is a natural coalgebra retract of $B(V)$. Thus $A^{\min}(V; M') \otimes A^{\min}(V; M'')$ is a natural coalgebra retract of $A^{\min}(V; M)$. So $A^{\min}(V; M') \otimes A^{\min}(V; M'')$ is also a natural coalgebra retract of $T(V)$ over $M(V) \cong M'(V) \oplus M''(V)$. By Corollary 4.13, $A^{\min}(V; M)$ is a natural coalgebra retract of $A^{\min}(V; M') \otimes A^{\min}(V; M'')$. The assertion follows. □

Corollary 6.9. *Suppose that there are functors $M_{n,\alpha}$ from $\mathbf{k}$-modules to $\mathbf{k}$-modules for $\alpha \in I_n$ and $n \geq 1$ such that there is a functorial isomorphism of $\mathbf{k}$-modules*

$$L_n^{\max}(V) \cong \bigoplus_{\alpha \in I_n} M_{n,\alpha}(V)$$

for $n \geq 1$. Then there is a functorial isomorphism of coalgebras

$$T(V) \cong \bigotimes_{n=1}^{\infty} \bigotimes_{\alpha \in I_n} A^{\min}(V; M_{n,\alpha}).$$

Let $M(V)$ be a subfunctor of $T(V)$ from **k**-modules to **k**-modules. The notations $A^{\min}(V;M)$ and $A^{\min}(M(V))$ have different meanings. The first one means the 'smallest' functorial coalgebra retract of $T(V)$ that contains $M(V)$ and the second one means the functor $A^{\min}(-)$ evaluated on the **k**-module $M(V)$ (thus the 'smallest' functorial retract of $T\big(M(V)\big)$ containing $M(V)$, where functorial means functorial in $M(V)$ rather than in V). A relation between $A^{\min}(V;M)$ and $A^{\min}(M(V))$ for some special functors M is as follows.

Proposition 6.10. *Let M be a subfunctor of L_n from* **k***-modules to* **k***-modules such that M is a retract of the functor $L_n^{\max}$. Then $A^{\min}(V;M)$ is a natural coalgebra retract of $A^{\min}(M(V))$ over $M(V)$.*

Proof. Let W denote $M(V)$. Notice that $A^{\min}(W)$ is a natural coalgebra retract of $T(W)$, where the inclusion and the retraction are functorial in W and so in V. By the proof of Proposition 6.8, $T(M(V))$ is a natural coalgebra retract of $T(V)$. Thus $A^{\min}(M(V))$ is a natural coalgebra retract of $T(V)$ over $M(V)$. The assertion follows from Corollary 4.13. □

$A^{\min}(V;M)$ can be a proper coalgebra retract of $A^{\min}\big(M(V)\big)$ as seen by the following example.

Example 6.11. Let the ground field **k** be of characteristic 2. Let $\phi\colon T(V)\to T(V)$ be the composite

$$T(V)\xrightarrow{H_6}T(V^{\otimes 6})\xrightarrow{T(\tau_{2,4})}T(V^{\otimes 6})\xrightarrow{T([\beta_3,\beta_3])}T([L_3(V),L_3(V)])\xhookrightarrow{j}T(V),$$

where j is the inclusion, $\tau_{2,4}(a_1a_2\cdots a_6)=a_1a_4a_3a_2a_5a_6$ given by interchanging positions 2 and 4 and

$$[\beta_3,\beta_3](a_1a_2\cdots a_6)=[[[a_1,a_2],a_3],[[a_4,a_5],a_6]].$$

Then we have

$$\begin{aligned}&\phi([[[x,y],y],[[y,z],z]]+[[[x,y],z],[[z,y],y]])\\&=([[[x,y],y],[[y,z],z]]+[[[x,y],z],[[z,y],y]])\end{aligned}$$

for $x,y,z\in V$. Let $B(V)=\operatorname{colim}_\phi T(V)$. Then $B_6(V)$ is non-trivial in general because ϕ has nonzero fixed point when $\dim(V)\geq 3$. By Corollary 4.4, $B(V)$ is a natural coalgebra retract of $T(V)$. Notice that the inclusion $T([L_3(V),L_3(V)])\to T(V)$ factors through the subHopf algebra $T(L_3(V))$ of $T(V)$. Therefore the map ϕ factors through $T(L_3(V))$. Thus $B(V)$ is a natural coalgebra retract of $T(L_3(V))$ and so there is a natural coalgebra decomposition

$$T(L_3(V))\cong B(V)\otimes A(V)$$

for some functor A from $\mathbf{k}$-modules to coalgebras. Notice that $B_j(V)=0$ for $0<j<6$ and so $L_3(V)\subseteq A(V)$. Thus $A^{\min}(V;L_3)$ is a natural retract of $A(V)$. Recall that $B_6(V)$ is non-trivial in general. Thus $A_6(V)$ is not functorially isomorphic to $L_3(V)\otimes L_3(V)$ and so $A_6^{\min}(V;L_3)$ is not functorially isomorphic to $L_3(V)\otimes L_3(V)$.

In contrast, we show that

$$A_2^{\min}(V)\cong T_2(V)=V\otimes V$$

for any $\mathbf{k}$-module V and so $A_6^{\min}(L_3(V))$ is functorially isomorphic to $L_3(V)\otimes L_3(V)$.

To check that $A_2^{\min}(V)$ is isomorphic to $V^{\otimes 2}$, let $j_V\colon A^{\min}(V)\to T(V)$ be the functorial inclusion and let $r_V\colon T(V)\to A^{\min}(V)$ be a functorial coalgebra retraction. Let $f_V=j_V\circ r_V\colon T(V)\to T(V)$ be the composite. Notice that f_V is a map of coalgebras and $f_V(x)=x$ for $x\in V$. There is an element $\zeta\in\mathbf{k}$ such that

$$f_V(xy)=(1+\zeta)xy-\zeta yx$$

for any $x,y\in V$. Notice that f_V is idempotent. Thus

$$f_V(xy)=f_V\circ f_V(xy)=((1+\zeta)^2+\zeta^2)xy-2\zeta(1+\zeta)yx=xy$$

for any $x,y\in V$ and so $A_2^{\min}(V)=\operatorname{colim}_{f_V}V^{\otimes 2}$ is isomorpic to $V^{\otimes 2}$.

If the field $\mathbf{k}$ is of characteristic 0, then the natural coalgebra decomposition of tensor algebras of Theorem 6.5 can be described explicitly. Let $S(V)$ denote the symmetric algebra generated by V.

Proposition 6.12. *If $\mathbf{k}$ is of characteristic 0, then there is functorial isomorphism of coalgebras*

$$A^{\min}(V;L_n^{\max})\cong S(L_n(V)).$$

Thus there is a functorial isomorphism of coalgebras

$$T(V)\cong\otimes_{n=1}^{\infty}S(L_n(V)).$$

Proof. First we show that $A^{\min}(V)$ is naturally isomorphic to $S(V)$.

Let $p\colon T(V)\to A^{\min}(V)$ be the canonical map induced by the multiplication in $A^{\min}(V)$. (See Proposition 6.1.) Let $B^{\max}(V)=\mathbf{k}\square_{A^{\min}(V)}T(V)$. By Proposition 6.1, $B^{\max}(V)$ is a sub-Hopf algebra of $T(V)$. Notice that $L_n(V)\subseteq B^{\max}(V)$ for $n\geq 2$. Let $\bar{B}(V)$ denote the sub-Hopf algebra of $T(V)$ generated by $L_n(V)$ with $n\geq 2$. Then $\bar{B}(V)$ is a normal sub-Hopf algebra of $T(V)$ with $\mathbf{k}\otimes_{\bar{B}(V)}T(V)\cong S(V)$ because all commutators are in $\bar{B}(V)$. Notice that the map $p\colon T(V)\to A^{\min}(V)$ factors through $\mathbf{k}\otimes_{\bar{B}(V)}T(V)\cong S(V)$. The resulting coalgebra map

$$\bar{p}\colon S(V)\to A^{\min}(V)$$

is an epimorphism. Notice that

$$\bar{p}\colon V=P(S(V))\to P(A^{\min}(V))=V$$

is an isomorphism. Thus $\bar{p}\colon S(V) \to A^{\min}(V)$ is a monomorphism and so $\bar{p}\colon S(V) \to A^{\min}(V)$ is an isomorphism of coalgebras. Since $L_1^{\max}(V) = V$, this shows the first statement for $n = 1$.

Now consider the case where $n \geq 2$. Notice that $L_n^{\max}(V) = L_n(V)$ according to Corollary 6.7. By the first step applied with $L_n(V)$ replacing V, there is an isomorphism

$$A^{\min}(L_n(V)) \cong S(L_n(V))$$

which is functorial in $L_n(V)$ and thus functorial in V. Therefore, by Proposition 6.10, $A^{\min}(V; L_n^{\max})$ is a natural coalgebra retract of $S(L_n(V))$ over $L_n(V)$. Consider the retraction map $r\colon S(L_n(V)) \to A^{\min}(V; L_n^{\max})$. Notice that

$$r\colon P(S(L_n(V))) = L_n(V) \to P(A^{\min}(V; L_n^{\max}))$$

is a monomorphism. Thus the retraction map $r\colon S(L_n(V)) \to A^{\min}(V; L_n^{\max})$ is a monomorphism. The assertion follows. □

7. Projective $\mathbf{k}(S_n)$-Submodules of Lie(n)

This section considers consequences of the earlier work in terms of modules over the group ring $\mathbf{k}(S_n)$ and in particular introduces the module $\mathrm{Lie}^{\max}(n)$. In the section we will sometimes work with an extension field $\mathbf{K}$ of the ground field $\mathbf{k}$. In this case, the tensor product of two $\mathbf{K}$-modules A and B over $\mathbf{K}$ will be denoted by $A \otimes_{\mathbf{K}} B$ and the n-th fold self tensor product of a $\mathbf{K}$-modules A over $\mathbf{K}$ will be denoted by $A^{\otimes_{\mathbf{K}} n}$.

Let $\bar{V}$ be the $\mathbf{k}$-module with basis $\{x_1, \ldots, x_n\}$ and let γ_n denote the sub $\mathbf{k}$-module of $\bar{V}^{\otimes n}$ generated by the elements $x_{\sigma(1)} \cdots x_{\sigma(n)}$, where σ runs through all elements in the symmetric group S_n. The $\mathbf{k}$-module Lie(n) is defined to be the intersection

$$\mathrm{Lie}(n) = \gamma_n \cap L_n(\bar{V}).$$

There are two $\mathbf{k}(S_n)$-actions on γ_n. One is induced by the canonical representation

$$\mathbf{k}(S_n) \to \mathrm{End}(\bar{V}),$$

and is called the internal $\mathbf{k}(S_n)$-action. Another (right) $\mathbf{k}(S_n)$-action on γ_n, called the position action, is given by permuting factors as described earlier. With these actions, γ_n becomes isomorphic as a $\mathbf{k}(S_n)$-bimodule to the group algebra $\mathbf{k}(S_n)$ itself. Given $\lambda \in \mathbf{k}(S_n)$ we also use λ to denote the $\mathbf{k}(S_n)$-module map $\gamma_n \to \gamma_n$ induced by the position action of λ. Observe that Lie(n) is the image of the map $\beta_n\colon \gamma_n \to \gamma_n$. Notice that the two $\mathbf{k}(S_n)$-actions on γ_n commute. It follows that β_n is a homomorphism with respect to the internal action. Thus the internal $\mathbf{k}(S_n)$-action on γ_n induces a $\mathbf{k}(S_n)$-action on $\mathrm{Lie}(n) = \mathrm{Im}\,\beta_n$. One can consider $\mathrm{Lie}(n)$ either as an internal $\mathbf{k}(S_n)$-submodule of γ_n using the inclusion

$$\mathrm{Lie}(n) \subseteq \gamma_n$$

or as an internal $\mathbf{k}(S_n)$ quotient module of γ_n under the identification

$$\mathrm{Lie}(n) = \mathrm{Im}(\beta_n \colon \gamma_n \to \gamma_n).$$

From here on, unless stated otherwise the phrase "$\mathbf{k}(S_n)$-module" will mean a module with respect to the *internal* action.
Note: $L_n(\bar{V})$ and $\mathrm{Lie}(n)$ are **not** closed under the position action.

Definition 7.1. *A* $\mathbf{k}(S_n)$*-module* P *is called a* **Lie projective module** *if*

(1). *P is a projective* $\mathbf{k}(S_n)$*-module;*
(2). *there exists a morphism of* $\mathbf{k}(S_n)$*-modules* $\phi\colon \gamma_n \to P$ *such that the composite*

$$\gamma_n \xrightarrow{\beta_n} \gamma_n \xrightarrow{\phi} P$$

is an epimorphism of $\mathbf{k}(S_n)$*-modules.*

We will show that P is a Lie projective module if and only if it is a projective $\mathbf{k}(S_n)$-module which is a submodule of $\mathrm{Lie}(n)$.

Lemma 7.2. *Let P be a* $\mathbf{k}(S_n)$*-module. Then P is a Lie projective module if and only if there exists an element $\lambda \in \mathbf{k}(S_n)$ such that there is an isomorphism of* $\mathbf{k}(S_n)$*-modules*

$$P \cong \mathrm{colim}_{\beta_n \circ \lambda}\, \gamma_n.$$

Proof. Suppose that P is a Lie projective module. Choose $\phi : \gamma_n \to P$ as in the definition. Let $s\colon P \to \gamma_n$ be a $\mathbf{k}(S_n)$-cross section map of the surjection

$$\gamma_n \xrightarrow{\beta_n} \gamma_n \xrightarrow{\phi} P.$$

Then

$$\mathrm{colim}_{\beta_n \circ (s \circ \phi)}\, \gamma_n \cong colim_{s\circ\phi\circ\beta_n} \gamma_n \cong \mathrm{Im}(s \circ \phi\colon \gamma_n \to \gamma_n) \cong P$$

and so one direction of the assertion follows. The converse is easy. □

Lemma 7.3. *Let P be any projective* $\mathbf{k}(S_n)$*-submodule of* $\mathrm{Lie}(n)$*. Then P is a Lie projective module.*

Proof. Let P be a projective $\mathbf{k}(S_n)$-submodule of $\mathrm{Lie}(n)$. Let $j\colon P \to \gamma_n$ denote the inclusion $P \subseteq \mathrm{Lie}(n) \subseteq \gamma_n$. Notice that $\mathbf{k}(S_n)$ is a Fröbenius algebra. (See [1, 10, 13].) Thus P is an injective $\mathbf{k}(S_n)$-module and so there exists a morphism of $\mathbf{k}(S_n)$-modules

$$r\colon \gamma_n \to P$$

such that the composite $r \circ j \colon P \to P$ is the identity map.

Let $\phi \colon \gamma_n \to \gamma_n$ denote the composite

$$\gamma_n \xrightarrow{\ r\ } P \hookrightarrow \mathrm{Lie}(n) \hookrightarrow \gamma_n.$$

Then $\phi(x_1 \cdots x_n) \in \mathrm{Lie}(n)$ and so there exists $k_\tau \in \mathbf{k}$ (for $\tau \in S_{n-1}$) such that

$$\phi(x_1 \cdots x_n) = \sum_{\tau \in S_{n-1}} k_\tau [[x_1, x_{\tau(2)}, \cdots, x_{\tau(n)}],$$

where S_{n-1} acts on $\{2, \dots, n\}$. Let $\lambda \colon \gamma_n \to \gamma_n$ be the morphism of internal $\mathbf{k}(S_n)$-modules determined by

$$\lambda(x_1 \cdots x_n) = \sum_{\tau \in S_{n-1}} k_\tau x_1 x_{\tau(2)} \cdots x_{\tau(n)}.$$

Then

$$\phi(x_1 \cdots x_n) = \beta_n \circ \lambda(x_1 \cdots x_n).$$

Since γ_n is the cyclic $\mathbf{k}(S_n)$-module generated by $x_1 \cdots x_n$ this implies $\phi = \beta_n \circ \lambda$. Notice that $\phi \circ \phi = \phi \colon \gamma_n \to \gamma_n$ and

$$P \cong \mathrm{colim}_\phi \, \gamma_n.$$

The assertion follows. □

Recall that there exists an element $\lambda_n \in \mathbf{k}(S_n)$ such that there is a natural isomorphism

$$\mathrm{colim}_{\beta_n \circ \lambda_n} V^{\otimes n} \cong L_n^{\max}(V)$$

for any $\mathbf{k}$-module V. (See Lemma 6.2.) Let $\mathrm{Lie}^{\max}(n)$ be defined as the colimit

$$\mathrm{Lie}^{\max}(n) \cong \mathrm{colim}_{\beta_n \circ \lambda_n} \gamma_n.$$

Notice that $\mathrm{Lie}^{\max}(n)$ is a Lie projective $\mathbf{k}(S_n)$-module.

Let P be a Lie projective $\mathbf{k}(S_n)$-module and let $\lambda \in \mathbf{k}(S_n)$ be such that

$$P \cong \mathrm{colim}_{\beta_n \circ \lambda} \gamma_n.$$

We may assume that P is a $\mathbf{k}(S_n)$-submodule of γ_n. Let V be any $\mathbf{k}$-module. Let the functor $L_n(-; P)$ be defined by

$$L_n(V; P) = \mathrm{colim}_{\beta_n \circ \lambda} V^{\otimes n}.$$

Theorem 7.4. *The $\mathbf{k}(S_n)$-module $\mathrm{Lie}^{\max}(n)$ is the maximum Lie projective $\mathbf{k}(S_n)$-module in the following sense:*

> *Let P be any Lie projective $\mathbf{k}(S_n)$-module. Then P is a $\mathbf{k}(S_n)$-retract of $\mathrm{Lie}^{\max}(n)$.*

Proof. Let $\bar{V}$ be the n-dimensional $\mathbf{k}$-module with basis $\{x_1, \ldots, x_n\}$. Let $d_j : \bar{V} \to \bar{V}$ be the map defined by

$$d_j(x_i) = \begin{cases} x_i & \text{if } i \neq j; \\ \\ 0 & \text{if } i = j \end{cases}$$

for $1 \leq j \leq n$. Notice that

$$\gamma_n = \cap_{j=1}^n \operatorname{Ker}(d_j : \bar{V}^{\otimes n} \to \bar{V}^{\otimes n}).$$

Thus there is an isomorphism of $\mathbf{k}(S_n)$-modules

$$P \cong \cap_{j=1}^n \operatorname{Ker}(L_n(- : P)(d_j) \colon L_n(\bar{V}; P) \to L_n(\bar{V}; P)).$$

The assertion follows from Proposition 6.6. □

Lemma 7.5. *Let* $\lambda, \mu \in \mathbf{k}(S_n)$. *Suppose that* $\operatorname{colim}_{\beta_n \circ \lambda} \gamma_n$ *is a* $\mathbf{k}(S_n)$*-retract of* $\operatorname{colim}_{\beta_n \circ \mu} \gamma_n$. *Then* $\operatorname{colim}_{\beta_n \circ \lambda} V^{\otimes n}$ *is a functorial retract of* $\operatorname{colim}_{\beta_n \circ \mu} V^{\otimes n}$ *for any* $\mathbf{k}$*-module* V.

Proof. By Corollary 4.4, one may assume that

(1). $\beta_n \circ \lambda \circ \beta_n \circ \lambda = \beta_n \circ \lambda \colon \gamma_n \to \gamma_n$.
(2). $\beta_n \circ \mu \circ \beta_n \circ \mu = \beta \circ \mu \colon \gamma_n \to \gamma_n$.

Let $M(V)$ and $N(V)$ denote the images

$$M(V) = \operatorname{Im}(\beta_n \circ \lambda \colon V^{\otimes n} \to V^{\otimes n}),$$

$$N(V) = \operatorname{Im}(\beta_n \circ \mu \colon V^{\otimes n} \to V^{\otimes n}).$$

Then $M(V)$ and $N(V)$ are functorially equivalent to $\operatorname{colim}_{\beta_n \circ \lambda} V^{\otimes n}$ and $\operatorname{colim}_{\beta_n \circ \mu} V^{\otimes n}$ respectively as $\mathbf{k}$-modules.

Let P and Q denote the images

$$P = \operatorname{Im}(\beta_n \circ \lambda \colon \gamma_n \to \gamma_n),$$

$$Q = \operatorname{Im}(\beta_n \circ \mu \colon \gamma_n \to \gamma_n).$$

Then P is a $\mathbf{k}(S_n)$-retract of Q. Let $f \colon P \to Q$ and let $g \colon Q \to P$ be morphisms of $\mathbf{k}(S_n)$-modules such that $g \circ f \colon P \to P$ is the identity map. Let $\theta(f), \theta(g) \colon \gamma_n \to \gamma_n$ denote the composites

$$\gamma_n \xrightarrow{r_P} P \xrightarrow{f} Q \hookrightarrow \gamma_n,$$

$$\gamma_n \xrightarrow{r_Q} Q \xrightarrow{g} P \hookrightarrow \gamma_n,$$

respectively, where $r_P\colon \gamma_n \to P$ and $r_Q\colon \gamma_n \to Q$ are retraction maps. The $\mathbf{k}(S_n)$-maps $\theta'(f), \theta'(g)\colon \gamma_n \to \gamma_n$ given respectively by the composites

$$\gamma_n \xrightarrow{\beta_n \circ \lambda} P \xrightarrow{f} Q \hookrightarrow \gamma_n,$$

and

$$\gamma_n \xrightarrow{\beta_n \circ \mu} Q \xrightarrow{g} P \hookrightarrow \gamma_n,$$

make the diagram

$$\begin{CD}
\gamma_n @>{\beta_n \circ \lambda}>> \gamma_n \\
@V{\theta'(f)}VV @VV{\theta(f)}V \\
\gamma_n @>{\beta_n \circ \mu}>> \gamma_n \\
@V{\theta'(g)}VV @VV{\theta(g)}V \\
\gamma_n @>{\beta_n \circ \lambda}>> \gamma_n
\end{CD}$$

commute. Let $\phi_f, \phi'_f, \phi_g, \phi'_g \in \mathbf{k}(S_n)$ denote the elements

$$\phi_f = \theta(f)(x_1 \cdots x_n), \phi'_f = \theta'(f)(x_1 \cdots x_n),$$

$$\phi_g = \theta(g)(x_1 \cdots x_n), \phi'_g = \theta'(g)(x_1 \cdots x_n).$$

The commutativity of the preceding diagram yields two equalities within $\mathbf{k}(S_n)$ which in turn imply that the diagram

$$\begin{CD}
V^{\otimes n} @>{\beta_n \circ \lambda}>> V^{\otimes n} \\
@V{\phi'_f}VV @VV{\phi_f}V \\
V^{\otimes n} @>{\beta_n \circ \mu}>> V^{\otimes n} \\
@V{\phi'_g}VV @VV{\phi_g}V \\
V^{\otimes n} @>{\beta_n \circ \lambda}>> V^{\otimes n}
\end{CD}$$

commutes naturally for any connected graded **k**-module V.

Notice that there is a commutative diagram

$$\begin{array}{ccc} \gamma_n & \xrightarrow{\beta_n \circ \lambda} & \gamma_n \\ \| & & \downarrow{\scriptstyle \theta(g)\circ\theta(f)} \\ \gamma_n & \xrightarrow{\beta_n \circ \lambda} & \gamma_n. \end{array}$$

Thus

$$\phi_g|_{N(V)} \circ \phi_f|_{M(V)} \colon M(V) \to M(V)$$

is the identity map. The assertion follows. □

Recall that the notation $\bar{V}$ means the n-dimensional **k**-module with basis $\{x_1, \ldots, x_n\}$. Let $d_j \colon \bar{V} \to \bar{V}$ be the j-th projection map. (See the proof of Theorem 7.4.) Let M be a functor from **k**-modules to **k**-modules. Let $\gamma(M) \subseteq M(\bar{V})$ be defined by

$$\gamma(M) = \bigcap_{j=1}^{n} \operatorname{Ker}(M(d_j) \colon M(\bar{V}) \to M(\bar{V})).$$

Notice that the (internal) $\mathbf{k}(S_n)$-action on $\bar{V}$ induces a $\mathbf{k}(S_n)$-action on $\gamma(M)$. The $\mathbf{k}(S_n)$-module $\gamma(M)$ is called the **associated $\mathbf{k}(S_n)$-module** of the functor M.

Proposition 7.6. *Let M be a subfunctor of the Lie functor L_n from **k**-modules to **k**-modules. Suppose that the associated $\mathbf{k}(S_n)$-module $\gamma(M)$ is a projective $\mathbf{k}(S_n)$-module. Then there exists an element $\lambda \in \mathbf{k}(S_n)$ such that:*

(1). $\beta_n \circ \lambda \circ \beta_n \circ \lambda = \beta_n \circ \lambda \colon V^{\otimes n} \to V^{\otimes n}$ *for any **k**-module V;*

(2). *there is a functorial isomorphism*

$$M(V) = \operatorname{Im}(\beta_n \circ \lambda \colon V^{\otimes n} \to V^{\otimes n}) \cong \operatorname{colim}_{\beta_n \circ \lambda} V^{\otimes n}.$$

Proof. Notice that

$$\gamma(M) = M(\bar{V}) \cap \gamma_n = M(\bar{V}) \cap \operatorname{Lie}(n).$$

By the proof of Lemma 7.3, there exists an element $\lambda \in \mathbf{k}(S_n)$ such that:

(1). $\beta_n \circ \lambda \circ \beta_n \circ \lambda = \beta_n \circ \lambda \colon V^{\otimes n} \to V^{\otimes n}$ for any connected graded **k**-module V;

(2). $\gamma(M) = \operatorname{Im}(\beta_n \circ \lambda \colon \gamma_n \to \gamma_n)$.

Let $\bar{M}$ be the functor from **k**-modules to **k**-modules defined by

$$\bar{M}(V) = \operatorname{Im}(\beta_n \circ \lambda \colon V^{\otimes n} \to V^{\otimes n})$$

for any **k**-module V. Notice that $\bar{M}(\bar{V}) \cap \gamma_n = M(\bar{V}) \cap \gamma_n$.

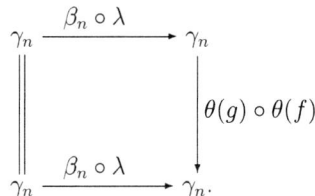

Let V be an $\mathbf{k}$-module and let $a_1, \ldots, a_n \in V$. Let $f\colon \bar{V} \to V$ be the morphism of $\mathbf{k}$-modules determined by

$$f(x_j) = a_j$$

for $1 \leq j \leq n$. Notice that

$$\beta_n \circ \lambda(x_1 \cdots x_n) \in \bar{M}(\bar{V}) \cap \gamma_n = M(\bar{V}) \cap \gamma_n.$$

Observe that there is a commutative diagram

$$\begin{array}{ccc} M(V) & \hookrightarrow & V^{\otimes n} \\ \Big\uparrow{\scriptstyle M(f)} & & \Big\uparrow{\scriptstyle f^{\otimes n}} \\ M(\bar{V}) & \hookrightarrow & \bar{V}^{\otimes n}. \end{array}$$

Thus

$$\beta_n \circ \lambda(a_1 \cdots a_n) = f^{\otimes n}(\beta_n \circ \lambda(x_1 \cdots x_n)) \in M(V)$$

for any $a_1, \ldots, a_n \in V$ and so one has a natural inclusion

$$\bar{M}(V) \subseteq M(V).$$

Now let $y \in M(V) \subseteq V^{\otimes n}$. Then there are elements $a_{ij} \in V$ with $1 \leq i \leq m$ and $1 \leq j \leq n$ such that

$$y = \sum_{i=1}^{m} k_i a_{i1} \cdots a_{in}$$

for some coefficients $k_1, \ldots, k_m \in \mathbf{k}$. Let $f_i\colon \bar{V} \to V$ be the morphism of $\mathbf{k}$-modules such that

$$f_i(x_j) = a_{ij}$$

for $1 \leq j \leq n$. Notice that there is a commutative diagram

$$\begin{array}{ccc} \bar{M}(V) & \longrightarrow & M(V) \\ \Big\uparrow{\scriptstyle \bar{M}(f_i)} & & \Big\uparrow{\scriptstyle M(f_i)} \\ \bar{M}(\bar{V}) & & M(\bar{V}) \\ \Big\uparrow\cup & & \Big\uparrow\cup \\ \bar{M}(\bar{V}) \cap \gamma_n & = & M(\bar{V}) \cap \gamma_n \end{array}$$

for each $1 \leq i \leq m$. Thus one has a commutative diagram

$$\begin{CD} \bar{M}(V) @>>> M(V) \\ @A{\sum_{i=1}^{m} \bar{M}(f_i)}AA @AA{\sum_{i=1}^{m} M(f_i)}A \\ \oplus_{i=1}^{m} \bar{M}(\bar{V}) \cap \gamma_n @= \oplus_{i=1}^{m} M(\bar{V}) \cap \gamma_n. \end{CD}$$

Notice that

$$y \in \operatorname{Im}(\sum_{i=1}^{m} M(\bar{V}) \cap \gamma_n \to M(V)).$$

Thus $y \in \bar{M}(V)$ and the assertion follows. □

Let $\mathbf{K}$ be an extension field of $\mathbf{k}$. For a $\mathbf{K}(S_n)$-module M the canonical inclusion $\mathbf{k}(S_n) \subseteq \mathbf{K}(S_n)$ induces a $\mathbf{k}(S_n)$-action on M, and we write $U(M)$ for the resulting $\mathbf{k}(S_n)$-module.

For a $\mathbf{k}$-module M we let $M^{\mathbf{K}}$ denote the $\mathbf{K}$-module $\mathbf{K} \otimes_{\mathbf{k}} M$. If M is a $\mathbf{k}(S_n)$-module, then M^K is a $\mathbf{K}(S_n)$-module. This gives a functor from the category of $\mathbf{k}(S_n)$-modules to the category of $\mathbf{K}(S_n)$-modules. The inclusion $\mathbf{k} \to \mathbf{K}$ induces a natural transformation

$$i \colon M \to U(M^{\mathbf{K}})$$

for $\mathbf{k}(S_n)$-modules M.

Let M be a $\mathbf{k}(S_n)$-module and let N be a $\mathbf{K}(S_n)$-module. Let $f \colon M \to U(N)$ be a morphism of $\mathbf{k}(S_n)$-modules. Let $e(f) \colon M^{\mathbf{K}} \to N$ be a morphism of $\mathbf{K}(S_n)$-modules defined by

$$f(a \otimes x) = af(x)$$

for $a \in K$ and $x \in M$. Thus for any $\mathbf{k}(S_n)$-module M and $\mathbf{K}(S_n)$-module N one has a natural function

$$e \colon \operatorname{Hom}_{\mathbf{k}(S_n)}(M, U(N)) \to \operatorname{Hom}_{\mathbf{K}(S_n)}(M^K, N)$$

which is an isomorphism of $\mathbf{K}$-modules. Thus, as is well known, the above associations form a pair of adjoint functors. (See, for example, [1, 11]).

Lemma 7.7. *Let $\mathbf{K}$ be an extension field of the ground field $\mathbf{k}$. Let M be a subfunctor of the Lie functor L_n from $\mathbf{k}$-modules to $\mathbf{k}$-modules. Suppose that $\mathbf{K} \otimes_{\mathbf{k}} \gamma(M)$ is a projective $\mathbf{K}(S_n)$-module. Then there exists an element $\lambda \in \mathbf{k}(S_n)$ such that*

(1). $\beta_n \circ \lambda \circ \beta_n \circ \lambda = \beta_n \circ \lambda \colon V^{\otimes n} \to V^{\otimes n}$ *for any connected graded $\mathbf{k}$-module V.*

(2). *there is a natural isomorphism*

$$M(V) = \mathrm{Im}(\beta_n \circ \lambda\colon V^{\otimes n} \to V^{\otimes n}) \cong \mathrm{colim}_{\beta_n \circ \lambda} V^{\otimes n}$$

for any $\mathbf{k}$*-module* V.

That is, $M(V)$ *is a Lie projective* $\mathbf{k}(S_n)$ *module.*

Proof. Notice that $\gamma(M)$ is a $\mathbf{k}(S_n)$-retract of $U(\mathbf{K} \otimes_{\mathbf{k}} \gamma(M))$. Notice that the functor U sends projective $\mathbf{K}(S_n)$-modules to projective $\mathbf{k}(S_n)$-modules. (See, for example, [11].) The assertion follows. □

Lemma 7.8. *Let* $\mathbf{K}$ *be an extension field of the ground field* $\mathbf{k}$*. Let*

$$f_V\colon T(V) \to T(V)$$

be a functorial map of coalgebras over the ground field $\mathbf{k}$*. Then there exists a functorial map of coalgebras over* $\mathbf{K}$*,* $\tilde{f}_W\colon T(W) \to T(W)$ *for* $\mathbf{K}$*-modules* W *such that* $\tilde{f}$ *is an extension of* f *in the following sense:*

Let V *be any* $\mathbf{k}$*-module. Then*

$$\tilde{f}_{V^{\mathbf{K}}} = \mathbf{K} \otimes_{\mathbf{k}} f\colon T(V^{\mathbf{K}}) = \mathbf{K} \otimes_{\mathbf{k}} T(V) \to T(V^{\mathbf{K}}) = \mathbf{K} \otimes_{\mathbf{k}} T(V).$$

Proof. There exists a sequence of elements $\alpha_n \in \mathbf{k}(S_n)$ for $n \geq 0$ such that

$$f_V = \alpha_n\colon T_n(V) = V^{\otimes n} \to T_n(V) = V^{\otimes n}.$$

Notice that $\alpha_n \in \mathbf{k}(S_n) \subseteq \mathbf{K}(S_n)$. Let W be any $\mathbf{K}$-module. Let

$$\tilde{f}_W\colon T(W) \to T(W)$$

be the functorial map defined by

$$\tilde{f}_W = \alpha_n\colon T_n(W) = W^{\otimes n} \to T_n(W) = W^{\otimes n}$$

for $n \geq 0$. Then one can check that:

1) $\tilde{f}$ is a functorial map of coalgebras over $\mathbf{K}$;

(1). $\tilde{f}$ is an extension of f.

The assertion follows. □

8. The Functor $A^{\min}$ over a Field of Characteristic $p > 0$

The purpose of this section is to obtain properties of the functor $A^{\min}$. In particular we prove the Cohen conjecture (Theorem 8.3) concerning the location of its primitives. Throughout the section, the ground field $\mathbf{k}$ is of characteristic $p > 0$.

8.1. An upper bound on the size of $A^{\min}(V)$.

Lemma 8.1.

Let $m>1$ such that $(m,p)=1$. Suppose that the polynomial

$$x^m-1$$

splits in $\mathbf{k}[x]$. Then there exists a functorial map of coalgebras

$$\phi_V\colon T(V)\to T(V)$$

for any $\mathbf{k}$-module V such that:

(1). $\phi_V\circ\phi_V=\phi_V\colon T(V)\to T(V)$;

(2). *Let $\alpha\in P(T_n(V))$ be a primitive element of tensor length n. Then*

$$\phi_V(\alpha)=\begin{cases}\alpha & \text{if} \quad n=m,2m,3m,\dots;\\ 0 & \text{otherwise.}\end{cases}$$

Proof. Let $\zeta\in\mathbf{k}$ be a primitive mth root of 1. Let

$$T(\zeta)\colon T(V)\to T(V)$$

be the algebra map determined by

$$T(\zeta)(v)=\zeta v$$

for $x\in V$. Then $T(\zeta)\colon T(V)\to T(V)$ is a functorial map of Hopf algebras. Let $\chi\colon T(V)\to T(V)$ be the conjugation and let

$$f_V=T(\zeta)*\chi\colon T(V)\to T(V)$$

be the convolution product. Then $f_V\colon T(V)\to T(V)$ is a functorial map of coalgebras.

Let $\alpha\in P(T_n(V))$ be a primitive element of tensor length n. Then

$$f_V(\alpha)=(\zeta^n-1)\alpha.$$

Let $D(V)$ be the colimit

$$D(V)=\operatorname{colim}_{f_V}T(V).$$

Let $r_V\colon T(V)\to D(V)$ be the canonical map to the colimit. By Theorem 4.5, $D(V)$ is a natural coalgebra retract of $T(V)$ and the functorial map r_V is a coalgebra retraction. Notice that $\zeta^n-1=0$ if and only if $n=mt$ for some $t\geq 0$. Thus

$$r_V\colon P(T_n(V))\to P(D_n(V))$$

is an isomorphism if n is not divisible by m and is zero if n is divisible by m.

By Lemma 5.2, there is a natural coalgebra decomposition

$$T(V)\cong B(V)\otimes D(V).$$

Let $r'_V\colon T(V)\to B(V)$ be a functorial coalgebra retraction and let $j_V\colon B(V)\to T(V)$ be the coalgebra injection. Let $\phi_V\colon T(V)\to T(V)$ be the composite

$$T(V)\xrightarrow{r'_V} B(V)\xrightarrow{j_V} T(V).$$

Then ϕ_V is an idempotent functorial coalgebra map. Let $\alpha\in P(T_n(V))$ be a primitive element of tensor length n. Notice that

$$P(T_n(V))\cong P(B_n(V))$$

if n is divisible by m and

$$P(B_n(V))=0$$

if n is not divisible by m. The assertion follows. □

Corollary 8.2. *Let $m>1$ such that $(m,p)=1$. Suppose that the*

$$x^m-1$$

splits in $\mathbf{k}[x]$. Let V be any $\mathbf{k}$-module. Then in tensor length mk we have

$$P(A^{\min}_{mk}(V))=0$$

for $k\geq 1$.

Proof. By Lemma 8.1, one has a natural coalgebra decomposition

$$T(V)\cong B(V)\otimes D(V)$$

such that $P(D_{mk}(V))=0$ for $k\geq 1$. Notice that there is a functorial inclusion

$$V\subseteq D(V).$$

Thus $A^{\min}(V)$ is a natural coalgebra retract of $D(V)$ by Theorem 4.12. The assertion follows. □

Theorem 8.3. *If the ground field $\mathbf{k}$ is of characteristic p, then*

$$P(A^{\min}_n(V))=0$$

if n is not a power of p.

Proof. Let m be a positive integer with $m>1$ and $(m,p)=1$. It suffices to show the following statement:

Statement: $P(A^{\min}_{mk}(V))=0$ for $k\geq 1$.

The proof of this statement will be given by induction on k. Let $k=1$. There is a natural coalgebra decomposition

$$T(V)\cong T(L_m(V))\otimes D(V)$$

for some natural subcoalgebra $D(V) \subseteq T(V)$. (See Corollary 6.7 or [9].) Notice that there is a functorial inclusion $V \subseteq D(V)$. Thus $A^{\min}(V)$ is a natural coalgebra retract of $D(V)$. Observe that the primitive submodule $P(T(V))$ is the free restricted Lie algebra $L^{\mathrm{res}}(V)$. Thus $P(T_m(V)) = L_m(V)$ and so $P(D_m(V)) = 0$. Therefore $P(A^{\min}_m(V)) = 0$.

Now suppose that
$$P(A^{\min}_{mt}(V)) = 0$$
for $t < k$ with $k > 1$ and let $n = mk$.

Goal 1: *The overall plan is to construct a natural coalgebra retract* $\overline{\overline{D}}(V)$ *such that* $V \subseteq \overline{\overline{D}}(V)$ *but* $P(\overline{\overline{D}}_{mk}(V)) = 0$.

By Theorem 6.5, there is a natural coalgebra decomposition
$$T(V) \cong \otimes_{j=2}^{\infty} A^{\min}(V; L_j^{\max}) \otimes A^{\min}(V).$$
Thus there is a functorial isomorphism of $\mathbf{k}$-modules
$$P(T(V)) \cong \oplus_{j=2}^{\infty} P(A^{\min}(V; L_j^{\max})) \oplus P(A^{\min}(V)).$$

Let
$$B(V) = \otimes_{j=2}^{n-1} A^{\min}(V; L_j^{\max})$$
be the partial tensor products in the natural decomposition for $T(V)$. By induction, $P(A^{\min}_{mt}(V)) = 0$ for $t < k$. Notice that $P(A^{\min}_{mt}(V; L_j^{\max})) = 0$ for $j \geq n = mk > mt$ by Corollary 6.4. Thus
$$P(T_{mt}(V)) \cong P(B_{mt}(V))$$
for $t < k$. Let $r_V : T(V) \to B(V)$ be a functorial coalgebra retraction, let $j_V : B(V) \to T(V)$ be a functorial coalgebra injection and let f_V denote the composite
$$f_V = j_V \circ r_V : T(V) \to T(V).$$
Then $f_V : T(V) \to T(V)$ is an idempotent functorial coalgebra map with a functorial isomorphism
$$\mathrm{Im}(f_V : T(V) \to T(V)) \cong B(V).$$
Goal 2: *The goal in the following is to change the ground field to one containing the* m*th roots of* 1.

Let $\mathbf{K}$ be an extension field of the ground field $\mathbf{k}$ such that the polynomial
$$x^m - 1$$
splits in $\mathbf{K}[x]$. By Lemma 7.8, there exists a functorial map of $\mathbf{K}$-coalgebras
$$\tilde{f}_W : T(W) \to T(W)$$

such that $\tilde{f}$ is an extension of f. Let $\tilde{B}(W)$ denote the colimit

$$\tilde{B}(W) = \operatorname{colim}_{\tilde{f}_W} T(W).$$

By Lemmas 5.2, 5.3 and Theorem 4.5, there is a natural $\mathbf{K}$-coalgebra decomposition

$$T(W) \cong \tilde{B}(W) \otimes_{\mathbf{K}} D(W)$$

for any finite dimensional $\mathbf{K}$-module W and so for any $\mathbf{K}$-module W, where $D(W)$ is given by the cotensor product

$$D(W) = \mathbf{K} \square_{\tilde{B}(W)} T(W).$$

Let $A^{\min,\mathbf{K}}$ denote the functor $A^{\min}$ over the extension field $\mathbf{K}$. Notice that there is a natural inclusion of $\mathbf{K}$-modules

$$W \subseteq D(W).$$

Thus $A^{\min,\mathbf{K}}(W)$ is a natural coalgebra retract of $D(W)$ and so one has a further natural $\mathbf{K}$-coalgebra decomposition

$$T(W) \cong \tilde{B}(W) \otimes D'(W) \otimes A^{\min,\mathbf{K}}(W)$$

for any $\mathbf{K}$-module W, where $D'(W)$ is given by the cotensor product

$$D'(W) = \mathbf{K} \square_{A^{\min,\mathbf{K}}(W)} D(W).$$

Goal 3: *The goal in the following is to give a further decomposition of $D'(W)$ by using Lemma 8.1.*

Let $r'_W \colon T(W) \to D'(W)$ be a functorial $\mathbf{K}$-coalgebra retraction and let $j'_W \colon D'(W) \to T(W)$ be a functorial $\mathbf{K}$-coalgebra injection. By Lemma 8.1, there exists an idempotent functorial map of $\mathbf{K}$-coalgebras

$$\phi_W \colon T(W) \to T(W)$$

such that the induced map on primitive elements

$$\phi_W \colon P(T_n(W)) \to P(T_n(W))$$

is the identity if $n = mt$ for some $t \geq 1$ and is zero if n is not divisible by m. Let the map $\phi'_W \colon T(W) \to T(W)$ denote the composite

$$T(W) \xrightarrow{r'_W} D'(W) \xrightarrow{j'_W} T(W) \xrightarrow{\phi_W} T(W).$$

Let $D''(W)$ denote the colimit

$$D''(W) = \operatorname{colim}_{\phi'_W} T(W).$$

Then $D''(W)$ is a natural $\mathbf{K}$-coalgebra retract of $D'(W)$. By Lemma 5.2, there is a natural $\mathbf{K}$-coalgebra decomposition

$$D'(W) \cong D''(W) \otimes D'''(W).$$

Notice that by Lemma 5.3

$$D''(W) \cong \mathbf{K} \square_{D'''(W)} D'(W)$$

for any finite dimensional $\mathbf{K}$-module W and so for any $\mathbf{K}$-module W. Thus we may view $D''(W) = \mathbf{K} \square_{D'''(W)} D'(W)$ as a subcoalgebra of

$$D'(W) \subseteq D(W) \subseteq T(W).$$

Consider the further natural $\mathbf{K}$-coalgebra decomposition

$$T(W) \cong \tilde{B}(W) \otimes D''(W) \otimes D'''(W) \otimes A^{\min,\mathbf{K}}(W).$$

There is a functorial isomorphism of $\mathbf{K}$-modules

$$P(T_l(W)) \cong P(\tilde{B}_l(W)) \oplus P(D_l''(W)) \oplus P(D_l'''(W)) \oplus P(A_l^{\min,\mathbf{K}}(W))$$

for any $l \geq 1$.

Goal 4: *The goal in the following is to show that $D_j''(W) = 0$ for $0 < j < mk$ and that $D_{mk}''(W)$ is a functorial $\mathbf{K}$-submodule of $L_{mk}^{\mathbf{K}}(W)$, where we write $L^{\mathbf{K}}$ for the free Lie algebra functor over the field $\mathbf{K}$.*

Now we determine the primitive elements $P(D_l''(W))$ up to tensor length $n = mk$. Let $\{x_i\}$ be a $\mathbf{K}$-basis for W and let W' be the $\mathbf{k}$-module generated by $\{x_i\}$. Then there is an isomorphism of $\mathbf{K}$-modules

$$\mathbf{K} \otimes_{\mathbf{k}} W' \cong W.$$

As noted earlier, there is an isomorphism of $\mathbf{k}$-modules

$$P(T_{mt}(W')) \cong P(B_{mt}(W'))$$

for $t < k$ and so, by Lemma 7.8, there is an isomorphism of $\mathbf{K}$-modules

$$P(T_{mt}(\mathbf{K} \otimes_{\mathbf{k}} W') \cong P(\tilde{B}_{mt}(\mathbf{K} \otimes_{\mathbf{k}} W')) = \mathbf{K} \otimes_{\mathbf{k}} P(B_{mt}(W'))$$

for $t < k$. In particular there is an isomorphism of $\mathbf{K}$-modules

$$P(T_{mt}(W)) \cong P(\tilde{B}_{mt}(W))$$

for $t < k$ and so

$$P(D_{mt}''(W)) \subseteq P(D_{mt}''(W)) \oplus P(D_{mt}'''(W)) \oplus P(A_{mt}^{min,\mathbf{K}}(W)) = 0$$

for $t < k$. Notice that $D''(W)$ is a natural $\mathbf{K}$-coalgebra retract of

$$\mathrm{Im}(\phi_W\colon T(W) \to T(W)) \cong \mathrm{colim}_{\phi_W} T(W).$$

Thus

$$P(D''_l(W)) \subseteq \mathrm{colim}_{\phi_W} P(T_l(W)) = 0$$

if l is not divisible by m. Therefore

$$P(D''_j(W)) = 0$$

for $j < mk$. Thus

$$D''_j(W) = 0$$

for $0 < j < mk$ and so one has

$$D''_{mk}(W) = P(D''_{mk}(W)).$$

Notice that the map

$$\phi_W\colon P(T_{mk}(W)) \to P(T_{mk}(W))$$

is the identity map. Thus there is a functorial isomorphism of $\mathbf{K}$-modules

$$P(D'_{mk}(W)) \cong P(D''_{mk}(W)) = D''_{mk}(W)$$

and so

$$P(D'''_{mk}(W)) = 0$$

for any $\mathbf{K}$-module W. By Corollary 8.2, one has

$$P(A^{\min,\mathbf{K}}_{mk}(W)) = 0$$

for any $\mathbf{K}$-module W. Thus there are functorial isomorphisms of $\mathbf{K}$-modules

$$P(T_{mk}(W)) \cong P(\tilde{B}_{mk}(W)) \oplus P(D''_{mk}(W)) = P(\tilde{B}_{mk}(W)) \oplus D''_{mk}(W);$$
$$P(D_{mk}(W)) = P(D'_{mk}(W)) = P(D''_{mk}(W)) = D''_{mk}(W).$$

Let $r''_W\colon T(W) \to D''(W)$ be a functorial $\mathbf{K}$-coalgebra retraction, let $j''_W\colon D''(W) \to T(W)$ be a functorial $\mathbf{K}$-coalgebra inclusion and let f''_W denote the composite

$$f''_W = j''_W \circ r''_W\colon T(W) \to T(W).$$

Notice that f''_W is a functorial map of $\mathbf{K}$-coalgebras with

$$f''_W = 0\colon T_j(W) \to T_j(W)$$

for $0 < j < mk$. By Corollary 2.9, there exists an element $\lambda \in \mathbf{K}(S_{mk})$ such that

$$f''_W|_{T_{mk}(W)} = \beta_{mk} \circ \lambda\colon T_{mk}(W) = W^{\otimes mk} \to T_{mk}(W) = W^{\otimes mk}.$$

Thus $D''_{mk}(W)$ is a functorial $\mathbf{K}$-submodule of $L^{\mathbf{K}}_{mk}(W)$ for any $\mathbf{K}$-module W, where we write $L^{\mathbf{K}}$ for the free Lie algebra functor over the field $\mathbf{K}$.

Goal 5: *The goal in the following is to construct the functor $\bar{\bar{D}}$ by using information about the functor D'' and to finish the proof.*

Now let V be any $\mathbf{k}$-module. Let $\bar{D}(V)$ denote the cotensor product

$$\bar{D}(V) = \mathbf{k}\square_{B(V)} T(V).$$

Let $W = \mathbf{K} \otimes_{\mathbf{k}} V$. Then

$$D(W) = \mathbf{K} \otimes_{\mathbf{k}} \bar{D}(V).$$

Thus $D''(W)$ is a subcoalgebra of $\mathbf{K} \otimes \bar{D}(V)$. Recall that

$$D''_{mk}(W) = P(D_{mk}(W)).$$

Thus

$$D''_{mk}(W) = \mathbf{K} \otimes_{\mathbf{k}} P(\bar{D}_{mk}(V)) \subseteq \mathbf{K} \otimes_{\mathbf{k}} L^{\mathrm{res}}_{mk}(V).$$

Recall that

$$D''_{mk}(W) \subseteq L^{\mathbf{K}}_{mk}(W) = \mathbf{K} \otimes_{\mathbf{k}} L_{mk}(V).$$

Thus there are no restricted elements in $P(\bar{D}_{mk}(V))$ or equivalently

$$P(\bar{D}_{mk}(V)) \subseteq L_{mk}(V).$$

Let $\gamma(P(\bar{D}_{mk}))$ be the associated $\mathbf{k}(S_{mk})$-module of the functor $P(\bar{D}_{mk})$. (See Proposition 7.6.) Notice that $D''_{mk}(W)$ is a functorial retract of $W^{\otimes n}$ for any $\mathbf{K}$-module W. Thus

$$\mathbf{K} \otimes_{\mathbf{k}} \gamma(P(\bar{D}_{mk}))) = \gamma(D''_{mk})$$

is a projective $\mathbf{K}(S_{mk})$-module. By Corollary 7.7, there exists an element $\bar{\lambda} \in \mathbf{k}(S_{mk})$ such that

(1). $\beta_{mk} \circ \bar{\lambda} \circ \beta_{mk} \circ \bar{\lambda} = \beta_{mk} \circ \bar{\lambda} \colon V^{\otimes mk} \to V^{\otimes mk}$ for any $\mathbf{k}$-module V;
(2). there is a functorial isomorphism of $\mathbf{k}$-modules

$$P(\bar{D}_{mk}(V)) = \mathrm{Im}(\beta_{mk} \circ \bar{\lambda} \colon V^{\otimes mk} \to V^{\otimes mk}) \cong \mathrm{colim}_{\beta_{mk} \circ \bar{\lambda}} V^{\otimes mk}$$

for any $\mathbf{k}$-module V.

We write $\bar{L}_{mk}$ for $P(\bar{D}_{mk})$. Then $A^{\min}(V; \bar{L}_{mk})$ is a natural coalgebra retract of $\bar{D}(V)$ and so there is a natural decomposition of connected graded coalgebras

$$\bar{D}(V) \cong A^{min}(V; \bar{L}_{mk}) \otimes \bar{\bar{D}}(V).$$

By Lemma 6.2, $\bar{L}_{mk}(V)$ is a natural retract of $L^{\max}_{mk}(V)$ and so $A^{\min}(V; \bar{L}_{mk})$ is a natural coalgebra retract of $A^{\min}(V; L^{\max}_{mk})$. By Corollary 6.4, we have

$$A^{\min}_j(V; \bar{L}_{mk}) = 0$$

for $0 < j < mk$, using Proposition 6.1 and Lemma 6.3. Thus there is a functorial inclusion $V \subseteq \overline{\overline{D}}(V)$ and so $A^{\min}(V)$ is a functorial retract of $\overline{\overline{D}}(V)$. Notice that

$$P(\bar{D}_{mk}(V)) = \bar{L}_{mk}(V) = P(A^{\min}_{mk}(V; \bar{L}_{mk})).$$

Thus $P(\overline{\overline{D}}_{mk}(V)) = 0$ and so

$$P(A^{\min}_{mk}(V)) = 0$$

for any **k**-module V. The induction is finished and the assertion follows. □

Recall that there is a functorial subHopf algebra $B^{\max}(V)$ such that there is a natural coalgebra decomposition $T(V) \cong B^{\max}(V) \otimes A^{\min}(V)$. (See Proposition 6.1.)

Corollary 8.4. *If n is not a power of p, then*

$$L_n(V) \subseteq B^{\max}(V).$$

Corollary 8.5. *$A^{\min}(V)$ is a natural coalgebra retract of the quotient of $T(V)$ modulo the right ideal generated by $\{L_n(V) \mid n$ not power of $p\}$.*

This gives an upper bound on the size of $A^{\min}$.

8.2. Some general theorems on natural coalgebra retracts of $T(V)$.

Theorem 8.6. *Let $B(V)$ be a functorial sub coalgebra of $T(V)$. Suppose that*

1) *There is a functorial multiplication $B(V) \otimes B(V) \to B(V)$ such that $B(V)$ is a quasi-Hopf algebra;*
2) *B is a retract of T as functors from **k**-modules to **k**modules.*

*Then B is a retract of T as functors from **k**-modules to **k**-coalgebras.*

Remark 8.7. *The inclusion $B(V)$ is only assumed to be a map of coalgebras. That is the inclusion might not be a map of algebras. Also we do not assume that the multiplication on $B(V)$ is associative.*

Proof. Let $B_m(V) = B(V) \cap T_m(V)$ and let $J_nB(V) = B(V) \cap J_n(V)$. Then $J_nB(V)$ is a functorial sub coalgebra of $J_n(V)$. Let $r\colon T(V) \to B(V)$ be a **k**-linear retraction. By Corollary 2.2, $r|_{J_n(V)}$ maps $J_n(V)$ onto $J_nB(V)$ and J_nB is a retract of J_n as functors from **k**-modules to **k**-modules. By Lemma 2.6, there is a functorial isomorphism of Hopf algebras $h_n\colon T(\bar{J}_n(V)) \to T(\bigoplus_{k=1}^n V^{\otimes k})$, where the Hopf algebra structure on $T(\bigoplus_{k=1}^n V^{\otimes k})$ is the one in which the elements in $\bigoplus_{k=1}^n V^{\otimes k}$ are primitive. Let $j_n\colon J_nB(V) \to J_n(V)$ be the inclusion and let $\bar{J}_nB(V) = J_nB(V)/J_0B(V)$. Let $\phi_n\colon T(\bar{J}_nB(V)) \to T(\bigoplus_{k=1}^n B_k(V))$ be the composite

$$T(\bar{J}_nB(V)) \xrightarrow{T(j_n)} T(\bar{J}_n(V)) \xrightarrow[\cong]{h_n} T(\bigoplus_{k=1}^{n} V^{\otimes k}) \xrightarrow{T(\oplus_{k=1}^n r|_{T_k(V)})} T(\bigoplus_{k=1}^{n} B_k(V)).$$

Then ϕ_n is a map of Hopf algebras for each n. By considering the commutative diagram of Hopf algebras

$$
\begin{array}{ccccccc}
T(B_n(V)) & \xrightarrow{T(j|_{B_n(V)})} & T(V^{\otimes n}) & = & T(V^{\otimes n}) & \xrightarrow{T(r|_{T_n(V)})} & T(B_n(V)) \\
\uparrow & & \uparrow & & \uparrow \text{proj} & & \uparrow \text{proj} \\
T(\bar{J}_n B(V)) & \hookrightarrow & T(\bar{J}_n(V)) & \underset{\cong}{\xrightarrow{h_n}} & T(\bigoplus_{k=1}^{n} V^{\otimes k}) & \twoheadrightarrow & T(\bigoplus_{k=1}^{n} B_k(V)) \\
\cup & & \cup & & \cup & & \cup \\
T(\bar{J}_{n-1} B(V)) & \hookrightarrow & T(\bar{J}_{n-1}(V)) & \underset{\cong}{\xrightarrow{h_{n-1}}} & T(\bigoplus_{k=1}^{n-1} V^{\otimes k}) & \twoheadrightarrow & T(\bigoplus_{k=1}^{n-1} B_k(V))
\end{array}
$$

we see that there is a commutative diagram of $\mathbf{k}$-modules

$$
\begin{array}{ccc}
B_n(V) & = & B_n(V) \\
\uparrow & & \uparrow \\
\bar{J}_n B(V) & \xrightarrow{Q(\phi_n)} & \bigoplus_{k=1}^{n} B_k(V) \\
\cup & & \cup \\
\bar{J}_{n-1} B(V) & \xrightarrow{Q(\phi_{n-1})} & \bigoplus_{k=1}^{n-1} B_k(V).
\end{array}
$$

Notice that $Q(\phi_1)\colon \bar{J}_1(B(V)) \to B_1(V)$ is an isomorphism. Thus $Q(\phi_n)\colon \bar{J}_n B(V) \to \bigoplus_{k=1}^{n} B_k(V)$ is an isomorphism by induction on n. It follows that $\phi_n\colon T(\bar{J}_n B(V)) \to T(\bigoplus_{k=1}^{n} B_k(V))$ is an epimorphism for each V. Notice that $T(\bar{J}_n B(V))$ has the same Poincaré series as $T(\bigoplus_{k=1}^{n} B_k(V))$ if $\dim(V) < \infty$. Thus ϕ_n is an isomorphism if $\dim(V) < \infty$ and so ϕ_n is an isomorphism for any $\mathbf{k}$-module V. Therefore the composite

$$
\phi\colon T(\overline{B(V)}) \xrightarrow{T(j)} T(\overline{T(V)}) \underset{\cong}{\xrightarrow{h}} T(\bigoplus_{n=1}^{\infty} V^{\otimes n}) \xrightarrow{T(r)} T(\bigoplus_{n=1}^{\infty} B_n(V))
$$

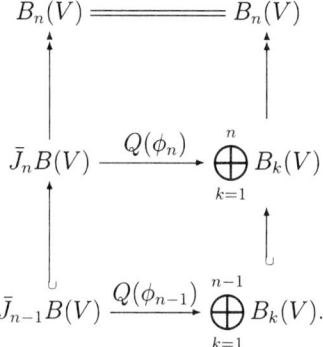

is an isomorphism of Hopf algebras, where $\overline{B(V)} = B(V)/B_0(V)$ and $\overline{T(V)} = T(V)/T_0(V)$.

By hypothesis $B(V)$ is a functorial quasi-Hopf algebra. Let $f_V : T(\overline{B(V)}) \to B(V)$ be defined by

$$f(x_1 x_2 \cdots x_n) = (\cdots(x_1 \cdot x_2)\cdots)\cdot x_n$$

for $x_j \in \overline{B(V)} = I(B(V))$. Then it is routine to check that f_V is a map of coalgebras. Let $E\colon B(V) \to T(\overline{B(V)})$ be the canonical inclusion. Notice that the composite

$$B(V) \overset{E}{\hookrightarrow} T(\overline{B(V)}) \underset{\cong}{\overset{\phi}{\longrightarrow}} T(\bigoplus_{n=1}^{\infty} B_n(V)) \underset{\cong}{\overset{\phi^{-1}}{\longrightarrow}} T(\overline{B(V)}) \overset{f_V}{\longrightarrow} B(V)$$

is the identity map of $B(V)$ and the composite $\phi \circ E\colon B(V) \to T(\bigoplus_{n=1}^{\infty} B_n(V))$ factors through $T(V)$. Thus $B(V)$ is a functorial coalgebra retract of $T(V)$, which is the assertion. □

If $B(V)$ is a functorial sub Hopf algebra of $T(V)$, we will show by example that there is no **functorial** isomorphism of Hopf algebras $T(Q(B(V)) \to B(V)$ although $B(V)$ is isomorphic to $T(Q(B(V))$ as Hopf algebras for any **individual** vector space V. However, for some special functorial sub Hopf algebras of $T(V)$, this statement holds.

Theorem 8.8. *Let $B(V)$ be a functorial sub Hopf algebra of $T(V)$. Suppose that B is a retract of T as functors from* **k***-modules to* **k***-modules. Then there is a functorial isomorphism of Hopf agebras*

$$T(Q(B(V)) \cong B(V).$$

Furthermore, $B(V)$ is generated by some elements in $L(V) \cap B(V)$.

Proof. Let $Q_n(V)$ denote $Q(B(V))_n$, the set of indecomposable elements of tensor length n. Notice that $B(V)$ is primitively generated and is a tensor algebra (being a subalgebra of a tensor algebra). It suffices to show that there is a functorial cross-section from $Q(B(V))$ to $L(V) \cap B(V) \subseteq P(B(V))$. This is given in assertion 1) of the following statement.

Statement: 1) There is a functorial submodule $Q'_n(V) \subseteq L_n(V) \cap B(V)$ such that the composite $Q'_n(V) \to L_n(V) \cap B(V) \overset{p_n}{\to} Q_n(V)$ is an isomorphism, where p_n is the restriction of the canonical map from $B_n(V)$ to $Q_n(V)$; and 2) $Q_n(V)$ is a functorial retract of $V^{\otimes n}$.

The proof of this statement is given by induction on n. The statement holds obviously for $n = 1$. Suppose that the statement holds for $k < n$ with $n > 1$.

Let $j_k\colon Q'_k(V) \to L_k(V) \cap B(V)$ be the inclusion for $k < n$ and let $j\colon \bigoplus_{k=1}^{n-1} Q'_k(V) \to \bigoplus_{k=1}^{n-1} L_k(V) \cap B(V) \subseteq B(V)$ be the inclusion induced by j_k. Let $T(j)\colon T(\bigoplus_{k=1}^{n-1} Q'_k(V))$

$\to B(V) \subseteq T(V)$ be the map of Hopf algebras induced by the map j. Notice that $B(V)$ is a tensor algebra and $j\colon \oplus_{k=1}^{n-1} Q'_k(V) \to Q(B(V))$ is a monomorphism. Thus $T(j)$ is a monomorphism. Notice that $Q_k(V)$ is a functorial retract of $V^{\otimes k}$ for $k < n$ and $Q'_k(V)$ is functorially isomorphic to $Q_k(V)$. Thus $Q'_k(V)$ is a functorial retract of $V^{\otimes k}$. Let $q \geq 1$ and let $k_1, k_2, \dots, k_q$ be positive integers with $k_j < n$. Let $l_1, l_2, \dots, l_q$ be positive integers. Then

$$Q_{k_1}(V)^{\otimes l_1} \otimes Q_{k_2}(V)^{\otimes l_2} \otimes \cdots \otimes Q_{k_q}(V)^{\otimes l_q}$$

is a functorial retract of $V^{\otimes k_1 l_1 + k_1 l_2 + \cdots k_q l_q}$. Let $m = k_1 l_1 + k_1 l_2 + \cdots k_q l_q$ and let $\bar{V} = \langle x_1, x_2, \dots, x_m \rangle$ be an m-dimensional vector space over $\mathbf{k}$. Let γ_m be the $\mathbf{k}$-submodule of $\bar{V}^{\otimes m}$ generated by the homogeneous elements $x_{\sigma(1)} x_{\sigma(2)} \cdots x_{\sigma(m)}$ for $\sigma \in S_m$. Then

$$(Q_{k_1}(V)^{\otimes l_1} \otimes Q_{k_2}(V)^{\otimes l_2} \otimes \cdots \otimes Q_{k_q}(V)^{\otimes l_q}) \cap \gamma_m$$

is a projective $\mathbf{k}(S_m)$-module. Notice that

$$T(\bigoplus_{k=1}^{n-1} Q'_k(V)) \cap T_m(V) = \bigoplus_{\substack{1 \leq k_1, k_2, \dots, k_q \leq n-1 \\ k_1 l_1 + k_2 l_2 + \cdots + k_q l_q = m}} Q_{k_1}(V)^{\otimes l_1} \otimes Q_{k_2}(V)^{\otimes l_2} \otimes \cdots \otimes Q_{k_q}(V)^{\otimes l_q}$$

for $m \geq 1$. Thus

$$T(\bigoplus_{k=1}^{n-1} Q'_k(\bar{V})) \cap T_m(\bar{V}) \cap \gamma_m$$

is a projective $\mathbf{k}(S_m)$-module. Notice that

$$T(\bigoplus_{k=1}^{n-1} Q'_k(V)) \cap T_m(V) = (T(\bigoplus_{k=1}^{n-1} Q'_k(\bar{V})) \cap T_m(\bar{V}) \cap \gamma_m) \otimes_{\mathbf{k}(S_m)} V^{\otimes m}$$

for any $\mathbf{k}$-module V. Thus

$$T(\bigoplus_{k=1}^{n-1} Q'_k(V)) \cap T_m(V)$$

is a functorial retract of $\gamma_m \otimes_{\mathbf{k}(S_m)} V^{\otimes m}$ and so $T(\bigoplus_{k=1}^{n-1} Q'_k(-))$ is a retract of T as functors from modules to modules. By Theorem 8.6, $T(\bigoplus_{k=1}^{n-1} Q'_k(V))$ is a functorial coalgebra retract of $T(V)$ and so $T(\bigoplus_{k=1}^{n-1} Q'_k(V))$ is a functorial coalgebra retract of $B(V)$. Let $r\colon B(V) \to T(\bigoplus_{k=1}^{n-1} Q'_k(V))$ be a functorial coalgebra retraction and let

$$B'(V) = \mathbf{k} \square_{T(\bigoplus_{k=1}^{n-1} Q'_k(V))} B(V).$$

By Lemma 5.3, there is a functorial isomorphism of coalgebras

$$B(V) \cong T(\bigoplus_{k=1}^{n-1} Q'_k(V)) \otimes B'(V).$$

Notice that $T(\bigoplus_{k=1}^{n-1} Q'_k(V))$ is a sub Hopf algebra of $B(V)$ generated by the elements $x \in B(V)$ with the tensor length $|x| < n$. Thus $B'(V)_j = 0$ for $0 < j < n$ and $B'(V)_n \cong Q_n(V)$. Let $Q'_n(V) = B'(V)_n$. Then $Q'_n(V) \subseteq P(B(V)) = L^{\mathrm{res}}(V) \cap B(V)$. Let $i\colon Q'_n(V) \to V^{\otimes n}$ be the inclusion of $Q'_n(V) \subseteq L^{\mathrm{res}}_n(V) \cap B(V) \subseteq V^{\otimes n}$. Notice that $B'(V)$ is a functorial retract of $T(V)$. Thus there is a functorial map $r'\colon V^{\otimes n} \to Q'_n(V)$ such that $r' \circ i\colon Q'_n(V) \to Q'_n(V)$ is the identity map and $Q'_n(V)$ is given by the set of the fixed points of the idempotent map $\phi = i \circ r'\colon V^{\otimes n} \to V^{\otimes n}$. Notice that $\mathrm{Im}(\phi\colon \gamma_n \to \gamma_n) \subseteq L^{\mathrm{res}}_n(\bar{V}) \cap \gamma_n = \mathrm{Lie}(n)$. Thus $\phi(x_1 \cdots x_n) \in L_n(\bar{V})$ and so $\phi(a_1 \cdots a_n) \in L_n(V)$ for any **k**-module V and any elements $a_j \in V$. Therefore $Q'_n(V) \subseteq L_n(V) \cap B(V)$. The induction is finished and the assertion follows. □

Let $Q^{\max}_n(V)$ denote $Q(B^{\max}(V))$ the set of indecomposable elements of $B^{\max}(V)$ with tensor length n.

Corollary 8.9. *The following statements hold:*

1) $Q^{\max}_1(V) = 0$;
2) $Q^{\max}_n$ *is a retract of* L_n *as functors from modules to modules for each* $n \geq 2$;
3) $Q^{\max}_n(V)$ *is a functorial retract of* $V^{\otimes n}$ *for each* $n \geq 2$;
4) *there is a functorial isomorphism of Hopf algebras*

$$T(\bigoplus_{n=2}^{\infty} Q^{\max}_n(V)) \cong B^{\max}(V).$$

A relation between the functor $Q^{\max}$ and the functor $A^{\min}$ is a follows.

Proposition 8.10. *There is a functorial isomorphism of* **k***-modules*

$$A^{\min}_{n-1}(V) \otimes V \cong Q^{\max}_n(V) \oplus A^{\min}_n(V).$$

for each $n \geq 1$.

Proof. Let $\tilde{A}^{\min}(V) = \mathbf{k} \otimes_{B^{\max}(V)} T(V)$. Then $\tilde{A}^{\min}(V)$ is functorially isomorphic to $A^{\min}(V)$ as a coalgebra. Let $p\colon T(V) \to \tilde{A}^{\min}(V)$ be the projection and let $I^{\max}(V)$ be the kernel of p. Notice that $I^{\max}(V)$ is a right ideal of $T(V)$ and $\tilde{A}^{\min}(V)$ is a right $T(V)$-module. Let $I\tilde{A}^{\min}(V)$ be the kernel of the augmentation map $\epsilon\colon \tilde{A}^{\min}(V) \to \mathbf{k}$ and let $q\colon T(V) \otimes V \to I\tilde{A}^{\min}(V)$ be the **k**-linear map given by

$$q(\alpha \otimes x) = p(\alpha x) = (\alpha) \cdot x$$

for $\alpha \in T(V)$ and $x \in V$, where $\cdot$ is the right action. Then

$$q(\alpha \otimes x) = 0$$

if $\alpha \in I^{\max}(V)$. Consider the commutative diagram of exact sequences of $\mathbf{k}$-modules

$$\begin{array}{ccccc}
I^{\max} \otimes V & = & I^{\max}(V) \otimes V & & \\
\downarrow{\scriptstyle \text{mult}} & & \downarrow & & \\
I^{\max}(V) & \hookrightarrow & T(V) \otimes V & \xrightarrow{q} & I\tilde{A}^{\min}(V) \\
\downarrow & & \downarrow{\scriptstyle p \otimes \mathrm{id}} & & \| \\
K(V) & \hookrightarrow & \tilde{A}^{\min}(V) \otimes V & \xrightarrow{\bar{q}} & I\tilde{A}^{\min}(V).
\end{array}$$

Then we have

$$K(V) \cong I^{\max}(V)/(I^{\max}(V) \cdot V) \cong I^{\max}(V)/(I^{\max}(V) \cdot IT(V)).$$

Notice that the canonical map $\pi\colon Q^{\max}(V) = IB^{\max}(V)/(IB^{\max}(V) \cdot IB^{\max}(V)) \to I^{\max}(V)/(I^{\max}(V) \cdot IT(V))$ is an epimorphism. The formula for $\chi(Q^{\max})$ in terms of $\chi(A^{\min})$ and $\chi(V)$ is determined by: 1) $T(Q^{\max}) \cong B^{\max}$ and 2) $T(V) \cong B^{\max} \otimes A^{\min}$. By comparing Poincaré series, we see that the map π is an isomorphism if $\dim(V) < \infty$ and so π is an isomorphism for any $\mathbf{k}$-module V. Thus $Q^{\max}(V) \cong K(V)$ and there is a short exact sequence

$$Q^{\max}_n(V) \hookrightarrow \tilde{A}^{\min}_{n-1}(V) \otimes V \twoheadrightarrow A^{\min}_n(V).$$

Notice that the $\mathbf{k}$-linear map $q\colon T(V) \otimes V \cong IT(V) \to IA^{\min}(V)$ has a functorial cross-section. The assertion follows. □

Remark 8.11. *It is the fact that $A^{\min}(V)$ and $\mathbf{k} \otimes_{B^{\max}(V)} T(V)$ are only functorially isomorphic as coalgebras rather than equal that prevents the filtration on $A^{\min}$ from being a Hopf algebra filtration instead of just a coalgebra filtration.*

8.3. A coalgebra filtration on the functor $A^{\min}$.

Now we give a coalgebra filtration on $A^{\min}(V)$ which can be used to give a lower bound on its growth. The filtration will be given by inductively defining a descending natural sequence of subcoalgebras of $A^{\min}(V)$.

Let $T^{(1)}(V) = T(V)$ and let $E^{(1)}(V) = \mathbf{k}[V]$ be the polynomial algebra generated by V. Let

$$T^{(2)}(V) = \mathbf{k}\,\square_{E^{(1)}(V)}\, T^{(1)}(V).$$

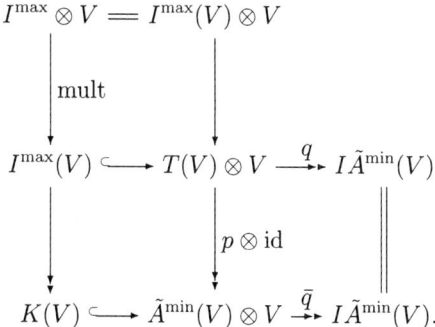

Notice that $T^{(2)}(V)$ is a normal subHopf algebra of $T(V)$ generated by $L_n(V)$ with $n \geq 2$. Thus $B^{\max}(V) \subseteq T^{(2)}$. Let $A^{(2)}(V)$ be defined by

$$A^{(2)}(V) = \mathbf{k} \otimes_{B^{\max}(V)} T^{(2)}(V).$$

Suppose that we have already defined $T^{(k)}(V)$ and $A^{(k)}(V)$ for $k \leq n$ and $E^{(k)}(V)$ for $k < n$ with the properties:

1) $T^{(k)}(V)$ is a subHopf algebra of $T(V)$ for $1 \leq k \leq n$ and
$$B^{\max}(V) \subseteq T^{(n)}(V) \subseteq T^{(n-1)}(V) \subseteq \cdots \subseteq T^{(2)}(V) \subseteq T^{(1)}(V) = T(V);$$
2) $E^{(k)}(V)$ is a commutative Hopf algebra such that
$$T^{(k+1)}(V) = \mathbf{k} \square_{E^{(k)}} T^{(k)}$$
for $k < n$.
3) $A^{(k)} = \mathbf{k} \otimes_{B^{\max}(V)} T^{(k)}(V)$.

We need to define $T^{(n+1)}(V)$, $A^{(n+1)}(V)$ and $E^{(n)}(V)$ for $n \geq 2$. Notice that $T^{(n+1)}(V)$ and $A^{(n+1)}(V)$ will be defined once $E^{(n)}(V)$ is defined. Now let $E^{(n)}(V)$ be defined to be the quotient Hopf algebra of $T^{(n)}(V)$ modulo the two sided ideal generated by

1) $B^{\max}(V)$;
2) the commutators $[x, y]$ for $x, y \in IT^{(n)}(V)$, where $IT^{(n)}(V)$ is the augmentation ideal of $T^{(n)}(V)$.

This gives a coalgebra filtration

$$\cdots A^{(n)}(V) \subseteq A^{(n-1)}(V) \subseteq \cdots \subseteq A^{(2)}(V) \subseteq A^{(1)}(V) \cong A^{\min}(V).$$

Observe that $\bigcap_{n=1}^{\infty} A^{(n)}(V) = 0$.

Proposition 8.12. *There is an equality of Poincaré series*

$$\chi(A^{\min}(V)) = \prod_{n=1}^{\infty} \chi(E^{(n)}(V)).$$

Proof. Notice that

$$\chi(T^{(n)}(V)) = \chi(T^{(n+1)}(V)) \cdot \chi(E^{(n)}(V)) = \chi(B^{\max}(V)) \cdot \chi(A^{(n)}(V));$$
$$\chi(T^{(n+1)}(V)) = \chi(B^{\max}(V)) \cdot \chi(A^{(n+1)}(V)).$$

Thus $\chi(A^{(n)}(V)) = \chi(A^{(n+1)}(V)) \cdot \chi(E^{(n)}(V))$. The assertion follows. □

Let $L(B^{\max}(V))$ denote $L(V) \cap B^{\max}(V)$. Let $L^{(1)}(V) = L(V)$ and let

$$L^{(n+1)}(V) = [L^{(n)}(V), L^{(n)}(V)] + L(B^{\max}(V))$$

be a sub Lie algebra of $L(V)$ for $n \geq 1$. This gives a descending sequence of Lie algebras

$$\cdots \subseteq L^{(n+1)}(V) \subseteq L^{(n)}(V) \subseteq \cdots \subseteq L^{(1)}(V) = L(V).$$

Let UL denote the universal enveloping algebra of a Lie algebra L.

Proposition 8.13. *Let $T^{(n)}$,$E^{(n)}$ and $L^{(n)}$ be defined as above. Then*

1) *there is a functorial isomorphism of Hopf algebras*

$$UL^{(n)}(V) \cong T^{(n)}(V)$$

for each $n \geq 1$;

2) *$E^{(n)}(V)$ is functorially isomorphic to the polynomial algebra generated by*

$$L^{(n)}(V)/L^{(n+1)}(V) = L^{(n)}(V)/([L^{(n)}(V), L^{(n)}(V)] + L(B^{\max}(V)))$$

for each $n \geq 1$.

Proof. Let $j_n\colon L^{(n)}(V) \to T^{(n)}(V)$ be the canonical inclusion. Notice that the composite $L^{(n)}(V) \longrightarrow T^{(n)}(V) \longrightarrow E^{(n)}(V)$ factors through $L^{(n)}(V)/L^{(n+1)}(V)$. Let $\bar{j}_n\colon L^{(n)}(V)/L^{(n+1)}(V) \longrightarrow E^{(n)}(V)$ be the induced map. We will show the following statements:

$\mathcal{P}_n$) The induced map of Hopf algebras

$$U(j_n)\colon UL^{(n)}(V) \to T^{(n)}(V)$$

is an isomorphism for each $n \geq 1$.

$\bar{\mathcal{P}}_n$) The induced map of Hopf algebras

$$U(\bar{j}_n)\colon U(L^{(n)}(V)/L^{(n+1)}(V)) \to E^{(n)}(V)$$

is an isomorphism for each $n \geq 1$.

Assertion 1) follows from statement $\mathcal{P}_n$. Notice that $L^{(n)}(V)/L^{(n+1)}(V)$ is an abelian Lie algebra and so assertion 2) follows from statement $\bar{\mathcal{P}}_n$. The proof of statements $\mathcal{P}_n$ and $\bar{\mathcal{P}}_n$ are given as follows.

1) Statement $\mathcal{P}_1$ holds obviously.
2) If statement $\mathcal{P}_n$ holds, then statement $\bar{\mathcal{P}}_n$ holds.
3) If statements $\mathcal{P}_n$ and $\bar{\mathcal{P}}_n$ hold, then statement $\mathcal{P}_{n+1}$ holds.

We first show that statement $\bar{\mathcal{P}}_n$ holds by assuming statement $\mathcal{P}_n$. By Corollary 8.9, $B^{\max}(V)$ is the sub Hopf algebra of $T(V)$ generated by $L(B^{\max}(V))$. By statement $\mathcal{P}_n$, $T^{(n)}(V)$ is the sub Hopf algebra of $T(V)$ generated by $L^{(n)}(V)$. Thus $E^{(n)}(V)$ is the quotient algebra of $T^{(n)}(V)$ modulo the two sided ideal generated by $L^{(n+1)}(V) = [L^{(n)}(V), L^{(n)}(V)] + L(B^{\max}(V))$. The standard arguments show that statement $\bar{\mathcal{P}}_n$ holds.

Now we show that statement $\bar{\mathcal{P}}_{n+1}$ holds by assuming statements $\mathcal{P}_n$ and $\bar{\mathcal{P}}_n$. It suffices to show statement $\mathcal{P}_{n+1}$ for the case that $\dim(V) < \infty$. Consider the short exact sequence of Lie algebras

$$L^{(n+1)}(V) \hookrightarrow L^{(n)}(V) \twoheadrightarrow L^{(n)}(V)/L^{(n+1)}(V).$$

There is a short exact sequence of Hopf algebras

$$UL^{(n+1)}(V) \hookrightarrow UL^{(n)}(V) \twoheadrightarrow U(L^{(n)}(V)/L^{(n+1)}(V)).$$

(See [5, Proposition 3.7].) Statement $\mathcal{P}_{n+1}$ follows from the following commutative diagram of Hopf algebras

$$\begin{array}{ccccc} UL^{(n+1)}(V) & \hookrightarrow & UL^{(n)}(V) & \twoheadrightarrow & U(L^{(n)}(V)/L^{(n+1)}(V)) \\ \downarrow & & \downarrow U(j_n) \cong & & \downarrow U(\bar{j}_n) \cong \\ T^{(n+1)}(V) & \hookrightarrow & T^{(n)}(V) & \twoheadrightarrow & E^{(n)}(V), \end{array}$$

where the rows are short exact sequences of Hopf algebras. This completes the proof. □

Notice that $A^{(n)}(V) = \mathbf{k} \otimes_{B^{\max}(V)} T^{(n)}(V)$ is a right $T^{(n)}(V)$-module. Let $IA^{(n)}(V)$ be the kernel of the augmentation map $A^{(n)}(V) \to \mathbf{k}$. By induction and using Proposition 8.13 and Corollary 8.4, we have

Corollary 8.14. *$E^{(n)}(V)$ is a polynomial algebra with indecomposable elements*

$$Q(E^{(n)}(V)) \cong IA^{(n)}(V) \otimes_{T^{(n)}(V)} \mathbf{k}.$$

Furthermore $Q(E^{(n)}(V))_j = 0$ for $j < p^{n-1}$.

Proposition 8.15. *If $\mathrm{char}(\mathbf{k}) > 2$, then $T^{(n+1)}(V)$ is the normal sub Hopf algebra of $T^{(n)}(V)$ generated by $B^{\max}(V)$.*

Proof. Let B' be the normal sub Hopf algebra of $T^{(n)}$ generated by $B^{\max}(V)$. Then $B' \subseteq T^{(n+1)}(V)$. By Proposition 8.13, it suffices to show that $L^{(n+1)}(V) \subseteq L(B') = B' \cap L(V)$. Notice that $L^{(n+1)}(V) = [L^{(n)}(V), L^{(n)}(V)] + L(B^{\max}(V))$. If q is not a power of $p = \mathrm{char}(\mathbf{k})$, then $L(B^{\max}(V))_q = L_q(V)$ by Corollary 8.4. Notice that $[L^{(n)}(V), L^{(n)}(V)]$ is a two sided Lie ideal generated by the elements $[\alpha, \beta]$ for $\alpha, \beta \in L^{(n)}(V)$. Also notice that $L(B')$ is a two sided Lie ideal of $L(T^{(n)}(V)) = T^{(n)}(V) \cap L(V)$. It follows that $[L^{(n)}(V), L^{(n)}(V)]_q \subseteq L(B')$ for any q. The assertion follows. □

8.4. A lower bound on the growth of $A^{\min}(V)$.

Now we give a lower bound for the functor $A^{\min}$ by determining $E^{(2)}(V)$. We need some preliminaries to determine $E^{(2)}(V)$.

Let S_n act on $\mathbf{k}^{\oplus n}$ by permuting the basis $\{y_1, y_2, \dots, y_n\}$ and let $I(\mathbf{k}^{\oplus n})$ be the kernel of the augmentation map $\epsilon\colon \mathbf{k}^{\oplus n} \to \mathbf{k}$ given by $\epsilon(y_j) = 1$ for $1 \le j \le n$. Then $I(\mathbf{k}^{\oplus n})$ is a sub $(\mathbf{S_n})$-module of $\mathbf{k}^{\oplus n}$. Let $Q(T^{(2)})_n$ be the set of indecomposable elements of $T^{(2)}(V)$ which have tensor length n and let S_n act on $V^{\otimes n}$ by permuting factors.

Proposition 8.16. *There is a functorial isomorphism of* $\mathbf{k}$*-modules*

$$Q(T^{(2)}(V))_n \cong I(\mathbf{k}^{\oplus n}) \otimes_{\mathbf{k}(S_n)} V^{\otimes n}$$

for any $\mathbf{k}$*-module* V.

Proof. By the proof of Lemma 3.13 in [5]. There is a functorial short exact sequence of $\mathbf{k}$-modules

$$Q(T^{(2)}) \hookrightarrow \mathbf{k}[V] \otimes V \xrightarrow{\text{mult}} I(\mathbf{k}[V]),$$

where $\mathbf{k}[V]$ is the polynomial algebra generated by V.

Let $\bar{V} = \langle x_1, \dots, x_n \rangle$ be an n-dimensional vector space over $\mathbf{k}$ and let γ_n be as defined earlier. Let M_n be the quotient $\mathbf{k}$-module of γ_n modulo the sub module generated by

$$x_{\sigma(1)} x_{\sigma(2)} \cdots x_{\sigma(i-1)} x_{\sigma(i+1)} \cdots x_{\sigma(n)} x_i - x_1 x_2 \cdots \hat{x}_i \cdots x_n x_i$$

for $1 \le i \le n$ and $\sigma \in S_{n-1}$ acting on $\{1, 2, \dots, \hat{i}, \dots, n\}$. Then M_n is a quotient $\mathbf{k}(S_n)$-module of γ_n. Let $q\colon \gamma_n \to M_n$ be the quotient map and let $y_i = q(x_1 \cdots \hat{x}_i \cdots x_n x_i) \in M_n$ for $1 \le i \le n$. Then $\{y_1, y_2, \dots, y_n\}$ is a $\mathbf{k}$-basis for M_n and there is an isomorphism of $\mathbf{k}(S_n)$-modules

$$M_n \cong \mathbf{k}^{\oplus n}.$$

Let $\epsilon_n\colon V^{\otimes n} \to \mathbf{k}[V]_n$ be the canonical quotient map. Then there is a commutative diagram

$$\begin{array}{ccccc}
V^{\otimes n} & \xrightarrow{\epsilon_{n-1} \otimes \mathrm{id}} & \mathbf{k}[V]_{n-1} \otimes V & \xrightarrow{\text{mult}} & \mathbf{k}[V]_n \\
\uparrow{\cong} & & \uparrow & & \uparrow{\cong} \\
\gamma_n \otimes_{\mathbf{k}(S_n)} V^{\otimes n} & \longrightarrow & M_n \otimes_{\mathbf{k}(S_n)} V^{\otimes n} & \xrightarrow{\epsilon \otimes \mathrm{id}} & \mathbf{k} \otimes_{\mathbf{k}(S_n)} V^{\otimes n},
\end{array}$$

where $\epsilon\colon M_n \to \mathbf{k}$ is given by $\epsilon(y_j) = 1$. Thus $M_n \otimes_{\mathbf{k}(S_n)} V^{\otimes n} \to \mathbf{k}[V]_{n-1} \otimes V$ is an epimorphism. Notice that $\dim(M_n \otimes_{\mathbf{k}(S_n)} V^{\otimes n}) \leq \dim(\mathbf{k}[V]_{n-1} \otimes V)$. Thus $M_n \otimes_{\mathbf{k}(S_n)} V^{\otimes n} \to \mathbf{k}[V]_{n-1} \otimes V$ is a (functorial) isomorphism. Consider the commutative diagram

$$\begin{array}{ccccc} Q(T^{(2)}(V))_n & \hookrightarrow & \mathbf{k}[V]_{n-1} \otimes V & \twoheadrightarrow & \mathbf{k}[V]_n \\ \uparrow & & \uparrow \cong & & \uparrow \cong \\ I(M_n) \otimes_{\mathbf{k}(S_n)} V^{\otimes n} & \to & M_n \otimes_{\mathbf{k}(S_n)} V^{\otimes n} & \twoheadrightarrow & \mathbf{k} \otimes_{\mathbf{k}(S_n)} V^{\otimes n}, \end{array}$$

where the top row is a short exact sequence.Thus $I(M_n) \otimes_{\mathbf{k}(S_n)} V^{\otimes n} \to Q(T^{(2)}(V))$ is an epimorphism. Notice that $\dim(I(M_n) \otimes_{\mathbf{k}(S_n)} V^{\otimes n}) \leq \dim(Q(T^{(2)}(V))$. Thus $I(M_n) \otimes_{\mathbf{k}(S_n)} V^{\otimes n} \to Q(T^{(2)}(V))$ is a functorial isomorphism if $\dim(V) < \infty$ and it is an isomorphism for any V, which is the assertion. □

Let $f_V\colon T(V) \to T(V)$ be a functorial map of coalgebras. Let $\alpha(f)_n \in \mathbf{k}(S_n)$ be the sequence of elements such that $\alpha(f)_n = f_V|_{T_n(V)}\colon T_n(V) = V^{\otimes n} \to T_n(V) = V^{\otimes n}$. (See section 7.) Let $\langle\ ,\ \rangle$ denote the canonical inner product in $\mathbf{k}(S_n)$. Notice that

$$\alpha = \sum_{\sigma \in S_n} \langle \alpha, \sigma \rangle \sigma$$

for any $\alpha \in \mathbf{k}(S_n)$.

Lemma 8.17. *Let $f_V\colon T(V) \to T(V)$ be a functorial map of coalgebras such that $(f_V)|_V\colon V \to V$ is the identity map for any V. Then*

$$\sum_{\sigma \in S_n} \langle \alpha(f)_n, \sigma \rangle = 1$$

for each $n \geq 1$.

Proof. Let $\bar{V}$ be the n-dimensional $\mathbf{k}$-module generated by $\{x_1, \ldots, x_n\}$. Let $q_1\colon T(V) \to T(V)$ be the projection map defined by

$$q_1(\alpha) = \begin{cases} \alpha & \text{if } \alpha \in V = T_1(V); \\ 0 & \text{if } \alpha \in T_n(V), n \neq 1. \end{cases}$$

Then one has

$$q_1 \circ f_V = q_1\colon T(\bar{V}) \to T(\bar{V}).$$

By hypothesis f is a map of coalgebras. Thus one has

$$q_1^{*n} \circ f = q_1^{*n}\colon T(\bar{V}) \to T(\bar{V}),$$

where q_1^{*n} is the n-fold self convolution product of q_1. Notice that

$$q_1^{*n}(x_1\cdots x_n)=\sum_{\sigma\in S_n}x_{\sigma(1)}\cdots x_{\sigma(n)}$$

and

$$\begin{aligned}q_1^{*n}\circ f(x_1\cdots x_n)&=\sum_{\sigma\in S_n}\langle\alpha(f)_n,\sigma\rangle q_1^{*n}(x_{\sigma(1)}\cdots x_{\sigma(n)})\\&=(\sum_{\sigma\in S_n}\langle\alpha(f)_n,\sigma\rangle)\sum_{\tau\in S_n}x_{\tau(1)}\cdots x_{\tau(n)}.\end{aligned}$$

The assertion follows. □

Corollary 8.18. *Let $f_V\colon T(V)\to T(V)$ be a natural map of coalgebras such that $(f_V)|_V\colon V\to V$ is the identity map for any V. Then*

(1). $f|_{T_n(V)}(x^n)=x^n$ *for any* $x\in V$ *and* $n\geq 1$;

(2).

$$f|_{T_n(V)}(\sum_{\sigma\in S_n}x_{\sigma(1)}\cdots x_{\sigma(n)})=\sum_{\sigma\in S_n}x_{\sigma(1)}\cdots x_{\sigma(n)}$$

for any $x_1,\dots x_n\in V$ *and* $n\geq 1$.

Lemma 8.19. *Let $f_V\colon T(V)\to T(V)$ be a natural map of coalgebras such that $(f_V)|_V\colon V\to V$ is the identity map for any V and let $x,y\in V$. Then*

$$f|_{T_{p^r}(V)}(ad^{p^r-1}(y)(x))=ad^{p^r-1}(y)(x)$$

for any $r\geq 1$.

Proof. We may assume that V is a two dimensional $\mathbf{k}$-module generated by x and y. The assertion follows from the formula:

$$(x+y)^{p^r}=y^{p^r}+ad^{p^r-1}(y)(x)+\Delta,$$

where Δ is a linear combination of the homogeneous elements in which x appears at least twice. (See [16].) □

Let $M_1,\dots,M_k$ be submodules of $T(V)$. We write $\langle M_1,\dots,M_k\rangle$ for the subHopf algebra of $T(V)$ generated by $M_1,\dots,M_k$. Given $n>1$, let $D^n(V)$ denote

$$\mathbf{k}\otimes_{\langle B^{max}(V),L_2,\dots,L_{n-1}\rangle}T(V),$$

where $L_j=L_j(V)$. Let $\alpha\in T(V)$. We write $\{\alpha\}$ for the image of α in $D^n(V)$.

Lemma 8.20. *Given $t\geq 1$, let $\bar{V}$ be the p^t-dimensional $\mathbf{k}$-module generated by $\{x_1,\dots,x_{p^t}\}$. Then the elements*

$$y_j=\{x_1\cdots\hat{x}_j\cdots x_{p^t}x_j\}$$

with $1\leq j\leq p^t$ are linearly independent in $D^{p^t}(\bar{V})$.

Proof. Suppose that $y_1, y_2, \dots, y_{p^t}$ are linearly dependent in $D^{p^t}(\bar{V})$. There exists $1 \le i \le p^t$ such that

$$y_i = \sum_{j \ne i} k_j y_j$$

for some $k_j \in \mathbf{k}$. Let $j \ne i$ with $1 \le j \le p^t$ and let $T_{i,j}\colon \bar{V} \cong \bar{V}$ be the $\mathbf{k}$-linear map given by $T_{i,j}(x_i) = x_j$,$T_{i,j}(x_j) = x_i$ and $T_{i,j}(x_k) = x_k$ for $k \ne i,j$. Let $D^{p^t}(T_{i,j})\colon D^{p^t}(\bar{V}) \cong D^{p^t}(\bar{V})$ be the induced isomorphism. Notice that

$$\{x_{i_1} \cdots x_{i_{p^t-1}} x_{i_{p^t}}\} = \{x_{i_{\tau(1)}} \cdots x_{i_{\tau(p^t-1)}} x_{i_{p^t}}\}$$

since $\langle B^{max}(V), L_2, \dots, L_{n-1}\rangle$ contains all commutators of length less than n. For any $\tau \in S_{p^t-1}$, one gets

$$D^{p^t}(T_{i,j})(y_k) = \begin{cases} y_j & \text{if} \quad k = i; \\ y_i & \text{if} \quad k = j; \\ y_k & \text{otherwise.} \end{cases}$$

Thus $y_j = \sum_{l \ne i,j} k_l y_l + k_j y_i$ and so $(1 + k_j)(y_i - y_j) = 0$. Let $\tilde{V}$ be the 2-dimensional $\mathbf{k}$-module generated by x and y and let $\phi\colon \bar{V} \to \tilde{V}$ be the $\mathbf{k}$-linear map given by $\phi(x_i) = x$ and $\phi(x_k) = y$ for $k \ne i$. Then

$$D^{p^t}(\phi)(y_i - y_j) = \{y^{p^t-1}x\} - \{xy^{p^t-1}\} = -\{ad^{p^t-1}(y)(x)\},$$

since we are in characteristic p. Consider the commutative diagram

$$\begin{array}{ccccc}
 & & D^{p^t}(\tilde{V}) & = = = = = & D^{p^t}(\tilde{V}) \\
 & & \uparrow & & \uparrow \\
B^{\max}(\tilde{V}) & \hookrightarrow & T(\tilde{V}) & \xrightarrow{r} & \mathbf{k} \otimes_{B^{\max}(\tilde{V})} T(\tilde{V}) \\
\| & & \uparrow & & \uparrow \\
B^{\max}(\tilde{V}) & \hookrightarrow & \langle B^{\max}, L_2, \dots, L_{p^t-1}\rangle & \twoheadrightarrow & \mathbf{k} \otimes_{B^{\max}(\tilde{V})} \langle B^{\max}(\tilde{V}), L_2, \dots, L_{p^t-1}\rangle.
\end{array}$$

Notice that $B^{\max}(\tilde{V})$ is a coalgebra retract of $T(\tilde{V})$. Thus $B^{\max}(\tilde{V})$ is a coalgebra retract of $\langle B^{\max}, L_2, \dots, L_{p^t-1}\rangle$ and so $\mathbf{k} \otimes_{B^{\max}(\tilde{V})} \langle B^{\max}, L_2, \dots, L_{p^t-1}\rangle$ is a coalgebra retract of $\langle B^{\max}, L_2, \dots, L_{p^t-1}\rangle$. Observe that $B_1^{\max}(V) = 0$, thus the composite

$$\langle L_2, \dots, L_{p^t-1}\rangle \hookrightarrow \langle B^{\max}(\tilde{V}), L_2, \dots, L_{p^t-1}\rangle \longrightarrow \mathbf{k} \otimes_{B^{\max}(\tilde{V})} \langle B^{\max}, L_2, \dots, L_{p^t-1}\rangle$$

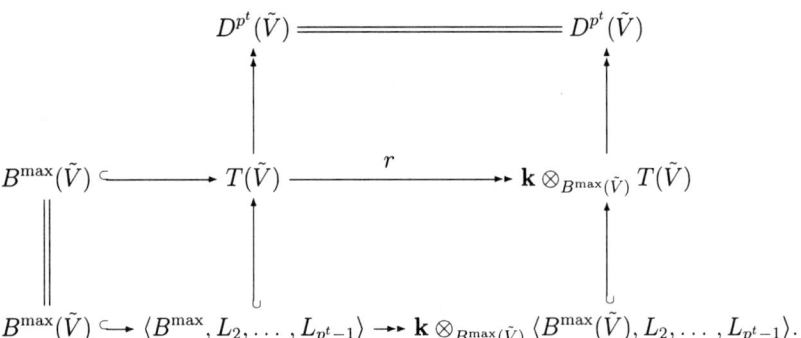

is onto up to dimension p^t. Notice that $r(ad^{p^t-1}(y)(x))$ is not zero by Lemma 8.19 and is not in the image of the composite $\langle L_2, \dots, L_{p^t-1}\rangle \to T(\tilde{V}) \to \mathbf{k} \otimes_{B^{\max}(\tilde{V})} T(\tilde{V})$. Thus the primitive element $r(ad^{p^t-1}(y)(x))$ is not in the subcoalgebra $\mathbf{k} \otimes_{B^{\max}(\tilde{V})} \langle B^{\max}(\tilde{V}), L_2, \dots, L_{p^t-1}\rangle$ and so

$$\{ad^{p^t-1}(y)(x)\} \neq 0$$

in $D^{p^t}(\tilde{V})$. Therefore $y_i - y_j \neq 0$ and so

$$k_j = -1$$

for each $j \neq i$. This shows that

$$y_1 + y_2 + \cdots + y_{p^t} = 0.$$

Notice that

$$D^{p^t}(\phi)(y_1 + \cdots + y_{p^t}) = \{y^{p^t-1}x\} + (p^t - 1)\{xy^{p^t-1}\} = -\{ad^{p^t-1}(y)(x)\} \neq 0.$$

One has a contradiction and the assertion follows. □

Let $q\colon T(V) \to D^{p^t}(V)$ be the quotient map and let $\gamma(D^{p^t}_{p^t})$ denote the image

$$\gamma(D^{p^t}_{p^t}) = \operatorname{Im}(q|_{\gamma_{p^t}} : \gamma_{p^t} \to D^{p^t}_{p^t}(\bar{V})).$$

Notice that $\gamma(D^{p^t}_{p^t})$ is a quotient $\mathbf{k}(S_{p^t})$-module of $\gamma_{p^t} \cong \mathbf{k}(S_{p^t})$. By Lemma 8.20, a representation of the $\mathbf{k}(S_n)$-module $\gamma(D^{p^t}_{p^t})$ is as follows.

Corollary 8.21. *There is an isomorphism of $\mathbf{k}(S_n)$-modules*

$$\gamma(D^{p^t}_{p^t}) \cong \mathbf{k}^{\oplus p^t},$$

where the $\mathbf{k}(S_{p^t})$-action on $\mathbf{k}^{\oplus p^t}$ is the standard representation.

Proposition 8.22. *Given $t \geq 1$ and let $\bar{V}$ be the p^t-dimensional $\mathbf{k}$-module generated by $\{x_1, \dots x_{p^t}\}$. Then the elements*

$$z_j = \{[[x_1, x_j, x_2, \dots, \hat{x}_j, \dots, x_{p^t}]\}$$

with $2 \leq j \leq p^t$ are linearly independent in $D^{p^t}(\bar{V})$.

Proof. The assertion follows from the fact

$$z_j = y_j - y_1$$

in $D^{p^t}(\bar{V})$. □

Let $\operatorname{Lie}(D^{p^t}_{p^t})$ denote the image

$$\operatorname{Lie}(D^{p^t}_{p^t}) = \operatorname{Im}(q|_{\operatorname{Lie}(p^t)} \colon \operatorname{Lie}(p^t) \to D^{p^t}_{p^t}(\bar{V})).$$

Corollary 8.23. *Let $a_1, \dots, a_{p^t}$ be the standard basis for $\mathbf{k}^{\oplus p^t}$. Then* $\mathrm{Lie}(D^{p^t}_{p^t})$ *is isomorphic to the* $\mathbf{k}$*-submodule of* $\mathbf{k}^{\oplus p^t}$ *spanned by* $a_2 - a_1, a_3 - a_1, \dots, a_{p^t} - a_1$ *as* $\mathbf{k}(S_{p^t})$*-modules.*

Proposition 8.24. *The* $\mathbf{k}$*-module* $D^n_j(V)$ *for* $j \le n$ *is determined as follows:*

$$D^n_j(V) \cong \begin{cases} \mathbf{k}[V]_j & if \quad j < n \\ \mathbf{k}[V]_n & if \quad j = n \neq p^t \ for\ some\ t \\ \mathbf{k}[V]_{p^t-1} \otimes V & if \quad j = n = p^t \end{cases}$$

Proof. Notice that $B^{\max}(V)$ is generated by certain choices of $L^{\max}_k(V) \subseteq L_k(V)$ for $k \geq 2$. Thus $\langle B^{\max}(V), L_2, \dots, L_{n-1}\rangle$ is equal to $\langle L_2, \dots, L_{n-1}, L^{\max}_n\rangle$ up to dimension n and so the only case we need to show is that

$$D^{p^t}_{p^t}(V) \cong \mathbf{k}[V]_{p^t-1} \otimes V.$$

Notice that the quotient map $T(V) \to D^{p^t}(V)$ induces a natural epimorphism

$$\theta\colon \mathbf{k}[V]_{p^t-1} \otimes V \to D^{p^t}_{p^t}(V).$$

By Lemma 8.20, it is routine to check that the natural map θ is a monomorphism. The assertion follows. □

Theorem 8.25. $E^{(2)}(V)$ *is functorially isomorphic to the polynomial algebra generated by*

$$\bigoplus_{t=1}^{\infty} I(\mathbf{k}^{\oplus p^t}) \otimes_{\mathbf{k}(S_{p^t})} V^{\otimes p^t}.$$

Proof. Let $t \geq 1$. Notice that $\langle B^{\max}(V), L_2, \dots, L_{p^t-1}\rangle \subseteq T^{(2)}(V)$. Let $D'(V) = \mathbf{k} \otimes_{\langle B^{\max}(V), L_2, \dots, L_{p^t-1}\rangle} T^{(2)}(V)$. Consider the commutative diagram

$$\begin{array}{ccccc}
\langle B^{\max}(V), L_2, \dots, L_{p^t-1}\rangle & \hookrightarrow & T^{(2)}(V) & \twoheadrightarrow & D'(V) \\
\| & & \cap\downarrow & & \cap\downarrow \\
\langle B^{\max}(V), L_2, \dots, L_{p^t-1}\rangle & \hookrightarrow & T(V) & \twoheadrightarrow & D^{p^t}(V) \\
 & & \downarrow & & \downarrow \\
 & & \mathbf{k}[V] & = & \mathbf{k}[V].
\end{array}$$

Notice that $D'(V)$ is $(p^t - 1)$-connected. By Proposition 8.24, there is a functorial isomorphism

$$D'(V)_{p^t} \cong \operatorname{Ker}(D^{p^t}_{p^t}(V) \to \mathbf{k}[V]_{p^t}) \cong \operatorname{Ker}(\mathbf{k}[V]_{p^t-1} \otimes V \to \mathbf{k}[V]_{p^t}).$$

By Proposition 8.16, the canonical map

$$Q(T^{(2)}(V))_{p^t} = (IT^{(2)}(V) \otimes_{T^{(2)}(V)} \mathbf{k})_{p^t} \to (I(D'(V)) \otimes_{T^{(2)}(V)} \mathbf{k})_{p^t} = D'(V)_{p^t}$$

is an isomorphism. Notice that the map $T^{(2)}(V) \to D'(V)$ factors through $A^{(2)}(V) = \mathbf{k} \otimes_{B^{\max}(V)} T^{(2)}(V)$. Thus the canonical map $Q(T^{(2)}(V))_{p^t} \to (I(A^{(2)}(V)) \otimes_{T^{(2)}(V)} \mathbf{k})_{p^t}$ is a monomorphism and so it is an isomorphism. By Corollary 8.14, the canonical map $Q(T^{(2)}(V))_{p^t} \to Q(E^{(2)})_{p^t}$ is an isomorphism. By Corollary 8.4, $Q(B^{\max}(V))_n \to Q(T^{(2)}(V))_n$ is onto if n is not a power of p and so the epimorphism $Q(T^{(2)}(V))_n \to Q(E^{(2)}(V))_n$ is the trivial map if n is not a power of p. The assertion follows. □

By Proposition 8.12, we have a lower bound on the growth of $A^{\min}(V)$.

Corollary 8.26. *If $\dim(V) \geq 2$, then $A^{\min}(V)$ has at least sub-exponential growth.*

Corollary 8.27. *There is a natural isomorphism*

$$A^{\min}_p(V) \cong \mathbf{k}[V]_{p-1} \otimes V.$$

Example 8.28. Let V be a two dimensional $\mathbf{k}$-module over the Steenrod algebra generated by $\{x, y\}$ with $P^1_*(y) = x$. Then, as a retract of $V^{\otimes p}$, $A^{\min}_p(V)$ is a $2p$-dimensional $\mathbf{k}$-module over Steenrod algebra with a basis:

$$\{y^p, P^1_*(y^p), \dots, P^p_*(y^p), z, P^1_*(z), \dots, P^{p-2}_*(z)\},$$

where $P^{p-1}_*(z) = P^p_*(y^p) = x^p$.

To conclude this section, as an example, we show that $T^{(2)}(V)$, which is the coalgebra kernel of the quotient map from tensor algebras to polynomial algebras, is not **functorially** isomorphic to the tensor algebra generated by $Q(T^{(2)}(V))$ if $\mathrm{char}(\mathbf{k}) > 0$.

Proposition 8.29. *The canonical quotient map $I(T^{(2)}(V)) \to Q(T^{(2)}(V))$ does NOT have a functorial $\mathbf{k}$-linear cross-section map.*

Proof. We show that $q\colon I(T^{(2)}(V))_n \to Q(T^{(2)}(V))_n$ does not have a functorial cross-section if $n \geq p+2$. Suppose that there were such a cross-section. Let $s\colon Q(T^{(2)}(V))_n \to I(T^{(2)}(V))_n$ be a functorial $\mathbf{k}$-linear map such that $q \circ s\colon Q(T^{(2)}(V))_n \to Q(T^{(2)}(V))_n$ is the identity. □

Let $\bar{V} = \langle x_1, x_2, \ldots, x_n\rangle$ be an n-dimensional vector space over $\mathbf{k}$ and let γ_n be as defined earlier. Let $M_n \cong \mathbf{k}^{\oplus n}$ be as defined in the proof of Proposition 8.16. Then there is a commutative diagram of short exact sequences of $\mathbf{k}(S_n)$-modules

$$\begin{array}{ccccc}
I(M_n) & \hookrightarrow & M_n & \xrightarrow{\ \epsilon\ } & \mathbf{k} \\
\uparrow{\scriptstyle p'} & & \uparrow{\scriptstyle p''} & & \| \\
I(\gamma_n) & \hookrightarrow & \gamma_n & \xrightarrow{\ \epsilon\ } & \mathbf{k},
\end{array}$$

where $\epsilon(x_{\sigma(1)} \cdots x_{\sigma(n)}) = 1$ for $\sigma \in S_n$.

Notice that $Q(T^{(2)}(V))_n$ is isomorphic functorially to $I(M_n) \otimes_{\mathbf{k}(S_n)} V^{\otimes n}$ by Proposition 8.16. Let $\phi\colon I(M_n) \to \bar{V}^{\otimes n}$ be the composite

$$I(M_n) = I(M_n) \otimes_{\mathbf{k}(S_n)} \gamma_n \subseteq I(M_n) \otimes_{\mathbf{k}(S_n)} \bar{V}^{\otimes n} \cong Q(T^{(2)}(\bar{V}))_n \xrightarrow{\ s\ } I(T^{(2)}(\bar{V}))_n \subseteq \bar{V}^{\otimes n}.$$

Then $\phi(I(M_n)) \subseteq I(T^{(2)}(\bar{V})) \cap \gamma_n \subseteq I(\gamma_n) \subseteq \gamma_n$ and the map $\phi\colon I(M_n) \to \gamma_n$ is a map of $\mathbf{k}(S_n)$ modules. Notice that γ_n is a free $\mathbf{k}(S_n)$-module. Therefore γ_n is an injective $\mathbf{k}(S_n)$-module (see [13]) and so there is a map of $\mathbf{k}(S_n)$-modules $\tilde{\phi}\colon M_n \to \gamma_n$ such that $\tilde{\phi}|_{I(M_n)} = \phi\colon I(M_n) \to \gamma_n$.

Let $f\colon \gamma_n \to \gamma_n$ be the composite

$$\gamma_n \xrightarrow{\ p''\ } M_n \xrightarrow{\ \tilde{\phi}\ } \gamma_n$$

and let $N = \mathrm{colim}_f\, \gamma_n$. Then N is a $\mathbf{k}(S_n)$-retract of γ_n and so N is a projective $\mathbf{k}(S_n)$-module. Recall that the composite $q \circ s\colon Q(T^{(2)}(V))_n \to Q(T^{(2)}(V))_n$ is the identity. It follows that the composite

$$I(M_n) \xrightarrow{\ \phi\ } I(T^{(2)}(\bar{V})) \cap \gamma_n \hookrightarrow I(\gamma_n) \xrightarrow{\ p'\ } I(M_n)$$

is the identity map and so

$$\dim(N) \geq \dim(I(M_n)) = n - 1.$$

On the other hand, N is a $\mathbf{k}(S_n)$-retract of M_n. Thus

$$\dim(N) \leq \dim(M_n) = n.$$

Now N is a projective $\mathbf{k}(S_n)$-module. Let P be the Sylow p-subgroup of S_n. Since projective modules over $\mathbf{k}(G)$ are free when G is a p-group, N is a free $\mathbf{k}(P)$-module and so $\dim(N)$ is divisible by the order of P. One gets a contradiction and the assertion follows.

9. Proof of Theorems 1.1 and 1.6

Let V^u be an arbitrary ungraded $\mathbf{k}$-module and let

$$T(V^u) \cong A(V^u) \otimes B(V^u)$$

be a natural coalgebra decomposition. Let $f_{V^u}, g_{V^u}\colon T(V^u) \to T(V^u)$ be the composites

$$T(V^u) \xrightarrow{r_A} A(V^u) \hookrightarrow T(V^u);$$

$$T(V^u) \xrightarrow{r_B} B(V^u) \hookrightarrow T(V^u),$$

respectively, where r_A and r_B are functorial coalgebra retractions. Let f^{grade} be as defined in Section 3. Let $A(V)$ and $B(V)$ be the colimits

$$A(V) = \operatorname{colim}_{f_V^{\mathrm{grade}}} T(V)$$

$$B(V) = \operatorname{colim}_{g_V^{\mathrm{grade}}} T(V)$$

for any connected graded $\mathbf{k}$-module V. Then the composite

$$T(V) \xrightarrow{\psi} T(V) \otimes T(V). \xrightarrow{\operatorname{colim}_{f_V^{\mathrm{grade}}} \otimes \operatorname{colim}_{g_V^{\mathrm{grade}}}} A(V) \otimes B(V)$$

is a functorial isomorphism by Lemma 5.2 for any finite dimensional $\mathbf{k}$-module V and so for any $\mathbf{k}$-module V. This extends the natural coalgebra decomposition of tensor algebras of ungraded $\mathbf{k}$-modules to the natural coalgebra decomposition of tensor algebra of connected graded $\mathbf{k}$-modules. We use the same notations A and B for the corresponding decomposition $T(V) \cong A(V) \otimes B(V)$. If $\mathbf{k} = \mathbb{Z}/p$, for any suspension X, by Theorem 1.3, there exist functorial maps $\phi_X, \phi'_X\colon \Omega\Sigma X \to \Omega\Sigma X$ such that

$$\phi_{X*} = f^{\mathrm{grade}}_{\bar{H}_*(X)}, \phi'_{X*} = g^{\mathrm{grade}}_{\bar{H}_*(X)}\colon H_*(\Omega\Sigma X) = T(\bar{H}_*(X)) \to H_*(\Omega\Sigma X) = T(\bar{H}_*(X)).$$

Let $A(X)$ and $B(X)$ be the homotopy colimits

$$A(X) = \operatorname{hocolim}_{\phi_X} \Omega\Sigma X$$
$$B(X) = \operatorname{hocolim}_{\phi'_X} \Omega\Sigma X$$

for any suspension X. Then $H_*(A(X)) = A(\bar{H}_*(X))$ and $H_*(B(X)) = B(\bar{H}_*(X))$. Let $\theta_X \colon \Omega\Sigma X \to A(X) \times B(X)$ be the composite

$$\Omega\Sigma X \xrightarrow{\Delta} \Omega\Sigma X \times \Omega\Sigma X \xrightarrow{\operatorname{hocolim}_{\phi_X} \times \operatorname{hocolim}_{\phi'_X}} A(X) \times B(X).$$

Then θ_X is a mod p homology equivalence and so θ_X is a homotopy equivalence if X is a p-torsion suspension of finite type. Notice that θ_X is functorial. Thus θ_X is a homotopy equivalence for any p-torsion suspension X. Again we use the same notation A and B for the corresponding decomposition $\Omega\Sigma X \simeq A(X) \times B(X)$.

Remark 9.1. *Observe that the functorial spaces $A(X)$ and $B(X)$ are well-defined with a functorial mod p homology equivalence $\Omega\Sigma X \xrightarrow{\simeq} A(X) \times B(X)$ for any suspension X. By considering the p-completion of the Cohen groups, one can show that $\Omega\Sigma X \simeq A(X) \times B(X)$ for any p-completed suspension X under certain choice of ϕ_X in Theorem 1.3.*

In the following proofs, the ground field $\mathbf{k} = \mathbb{Z}/p$.

Proof of Theorem 1.1. By the natural coalgebra decomposition

$$T(V^u) \cong A^{\min}(V^u) \otimes B^{\max}(V^u),$$

one gets a natural decomposition

$$\Omega\Sigma X \simeq A^{\min}(X) \times B^{\max}(X)$$

with $H_*(X) \subseteq H_*(A(X))$ for any p-torsion suspension X.

Let $f_{V^u} \colon T(V^u) \to T(V^u)$ be the composite

$$T(V^u) \xrightarrow{r_{A^{\min}}} A^{\min}(V^u) \hookrightarrow T(V^u).$$

By Corollary 8.4, $f_{V^u}|_{T_n(V^u)} \circ \beta_n \colon (V^u)^{\otimes n} \to (V^u)^{\otimes n}$ is zero if n is not a power of p, where $\beta_n(x_1 \cdots x_n) = [[x_1, x_2], \cdots, x_n]$, the n-fold (graded or ungraded) commutator. Thus $f_V^{\mathrm{grade}}|_{T_n(V)} \circ \beta_n \colon V^{\otimes n} \to V^{\otimes n}$ is zero if n is not a power of p and so $L_n(V) \subseteq B^{\max}(V)$ for any connected graded $\mathbf{k}$-module V for such n. In particular, $L_n(\bar{H}_*(X)) \subseteq H_*(B^{\max}(X))$ if n is not a power of p.

We show that $H_*(B^{\max}(X))$ is a subHopf algebra of $H_*(\Omega\Sigma X)$. It suffices to show that $B^{\max}(V)$ is a subHopf algebra of $T(V)$ for any connected graded $\mathbf{k}$-module V.

Let $g_{V^u} \colon T(V^u) \to T(V^u)$ be the composite

$$T(V^u) \xrightarrow{r_{B^{\max}}} B^{\max}(V^u) \hookrightarrow T(V^u).$$

By Lemma 2.1, there is a sequence of elements $\alpha(g)_n \in \mathbf{k}(S_n)$ such that

$$\alpha(g)_n = g_{V^u}|_{T_n(V^u)} \colon T_n(V^u) \to T_n(V^u).$$

Notice that $B^{\max}(V^u)$ is a subHopf algebra of $T(V^u)$. There exists $k_\sigma \in \mathbf{k}$ for $\sigma \in S_{m+n}$ such that

$$(1) \quad \alpha(g)_m(x_1 \cdots x_m) \cdot \alpha(g)_n(x_{m+1} \cdots x_{m+n}) = \sum_{\sigma \in S_{m+n}} k_\sigma \alpha(g)_{n+m}(x_{\sigma(1)} \cdots x_{\sigma(m+n)})$$

for $m, n \geq 0$ and arbitrary $x_1, \ldots, x_{m+n}$.

Let V be a connected graded $\mathbf{k}$-module. Notice that $g_V^{\mathrm{grade}}|_{T_n(V)} \colon T_n(V) \to T_n(V)$ is given by

$$g_V^{\mathrm{grade}}|_{T_n(V)} = \alpha(g)_n \otimes id \colon V^{\otimes n} = \gamma_n \otimes_{\mathbf{k}(S_n)} V^{\otimes n} \to V^{\otimes n} = \gamma_n \otimes_{\mathbf{k}(S_n)} V^{\otimes n},$$

where S_n acts internally on γ_n (see Section 7) and S_n acts on $V^{\otimes n}$ by permuting factors in **graded** sense. Notice that $B^{\max}(V)$ is the image of the idempotent map $g_V^{\mathrm{grade}} \colon T(V) \to T(V)$. Thus $B^{\max}(V)$ is closed under the multiplication by (refeq:1) above and so $B^{\max}(V)$ is a sub Hopf algebra of $T(V)$.

Now we show that $B^{\max}(X)$ is a loop suspension and that the injection $B^{\max}(X) \to \Omega\Sigma X$ can be chosen as a loop map. By Corollary 8.9, there is a functorial sub $\mathbf{k}$-module $Q_n^{\max}(V^u)$ of $L_n(V^u) \cap B^{\max}(V^u)$ such that

1) $Q_n^{\max}(V^u)$ is a functorial retract of $V^{u \otimes n}$;
2) The canonical map $T(\oplus_{n=2}^{\infty} Q_n^{\max}(V^u)) \to B^{\max}(V^u)$ is a functorial isomorphism. That is $Q_n^{\max}(V^u)$ with $n \geq 2$ are algebraically independent and $B^{\max}(V^u)$ is a sub Hopf algebra of $T(V^u)$ generated by $Q_n^{\max}(V^u)$ for $n \geq 2$.

By Proposition 7.6, there exists an element $\lambda_n \in \mathbf{k}(S_n)$ such that

$$\beta_n \circ \lambda_n \circ \beta_n \circ \lambda_n = \beta_n \circ \lambda_n$$

and $Q_n^{\max}(V^u) = \mathrm{colim}_{\beta_n \circ \lambda_n} V^{u \otimes n}$. Let V be any grade $\mathbf{k}$-module. Let $Q_n^{\max}(V) \subseteq L_n(V)$ be defined by

$$Q_n^{\max}(V) = \mathrm{Im}(\beta_n \circ \lambda_n \colon V^{\otimes n} \to V^{\otimes n}) \cong \mathrm{colim}_{\beta_n \circ \lambda_n} V^{\otimes n},$$

where the map $\alpha \colon V^{\otimes n} \to V^{\otimes n}$ is defined by the S_n (graded) action on $V^{\otimes n}$ for any $\alpha \in \mathbf{k}(S_n)$. Then $B^{\max}(V)$ is a sub Hopf algebra of $T(V)$ generated by $Q_n^{\max}(V)$ for $n \geq 2$ and the canonical map $T(\oplus_{n=2}^{\infty} Q_n^{\max}(V)) \to B^{\max}(V)$ is a functorial isomorphism of Hopf algebras. Now let X be a suspension. Let S_n act on the n-fold self smash product $X^{(n)}$ by permuting coordinates. Notice that $[X^{(n)}, X^{(n)}]$ is an abelian group. The map $\alpha \colon X^{(n)} \to X^{(n)}$ is well defined by using S_n-action for any

$\alpha \in \mathbb{Z}(S_n)$. Let $q\colon \mathbb{Z}(S_n) \to \mathbb{Z}/p(S_n)$ be the quotient map and let $\bar{\lambda}_n \in \mathbb{Z}(S_n)$ such that $q(\bar{\lambda}_n) = \lambda_n$ for each $n \geq 2$. Let $Q_n^{\max}(X)$ be defined by the homotopy colimit

$$Q_n^{\max}(X) = \operatorname{hocolim}_{\beta_n \circ \bar{\lambda}_n} X^{(n)}.$$

Notice that

$$(\beta_n \circ \bar{\lambda}_n)_* = \beta_n \circ \lambda_n\colon \bar{H}_*(X^{(n)}) = (\bar{H}_*(X))^{\otimes n} \to \bar{H}_*(X^{(n)}) = (\bar{H}_*(X))^{\otimes n}$$

is an idempotent map. Thus the composite

$$X^{(n)} \xrightarrow{\text{comult}} X^{(n)} \vee X^{(n)} \longrightarrow \operatorname{hocolim}_{\mathrm{id}-\beta_n \circ \bar{\lambda}_n} X^{(n)} \vee \operatorname{hocolim}_{\beta_n \circ \bar{\lambda}_n} X^{(n)}$$

is a mod p homology equivalence and so $Q_n^{\max}(X)$ is a functorial retract of $X^{(n)}$ for any p-torsion suspension X. Let $p_m\colon X^{(n)} \to Q_n^{\max}(X)$ be the canonical map from the m-th term in the sequence

$$X^{(n)} \xrightarrow{\beta_n \circ \bar{\lambda}_n} X^{(n)} \xrightarrow{\beta_n \circ \bar{\lambda}_n} \cdots$$

to the homotopy colimit. Let $s\colon Q_n^{\max}(X) \to X^{(n)}$ be a functorial map such that $p_1 \circ s = \mathrm{id}_{Q_n^{\max}(X)}$. Let $j_n\colon Q_n^{\max}(X) \to J(X)$ be the composite □

$$Q_n^{\max}(X) \xrightarrow{s} X^{(n)} \xrightarrow{\bar{\lambda}_n} X^{(n)} \xrightarrow{W_n} J(X),$$

where W_n is the iteratated (left to right) Samelson product. Let $J(j_n)$: $J(Q_n^{\max}(X)) \to J(X)$ be the map of topological monoids induced by the map j_n. Let $H_n\colon J(X) \to J(X^{(n)})$ be the James-Hopf map. Notice that $H_n \circ J(W_n) \simeq J(\beta_n)\colon J(X^{(n)}) \to J(X^{(n)})$. (See [17].) Thus $J(p_2) \circ H_n \circ J(j_n)\colon J(Q_n^{\max}(X)) \to J(Q_n^{\max}(X))$ is homotopic to the identity map and so $j_{n*}\colon \bar{H}_*(Q_n^{\max}(X)) \to H_*(J(X)) = T(\bar{H}_*(X))$ maps onto the submodule

$$\operatorname{Im}((W_n \circ \bar{\lambda}_n)_*\colon \bar{H}_*(X^{(n)}) \to \bar{H}_*(J(X))) = \operatorname{Im}(\beta_n \circ \lambda_n\colon V^{\otimes n} \to V^{\otimes n}) = Q_n^{\max}(V),$$

where $V = \bar{H}_*(X)$. Let $Q^{\max}(X) = \bigvee_{n=2}^{\infty} Q_n^{\max}(X)$ and let $j\colon Q^{\max}(X) \to J(X)$ be the map such that $j|_{Q_n^{\max}(X)} = j_n$: $Q_n^{\max}(X) \to J(X)$. Let $J(j)$: $J(Q^{\max}(X)) \to J(X)$ be the map of topological monoids induced by the map j. Then $J(j)_*$: $H_*(J(Q^{\max}(X))) \to H_*(J(X))$ maps onto $B^{\max}(\bar{H}_*(X))$ which is the sub Hopf algebra generated by $Q_n^{\max}(\bar{H}_*(X))$. Let $r\colon J(X) \to B^{\max}(X)$ be a retraction. Then the composite

$$(r \circ J(j))_*\colon H_*(J(Q^{\max}(X))) \to H_*(B^{\max}(X)) = B^{\max}(\bar{H}_*(X))$$

is an epimorphism. Notice that $H_*(J(Q^{\max}(X)))$ and $H_*(B^{\max}(X))$ have the same Poincaré series if $H_*(X)$ is of finite type. Thus $r \circ J(j)\colon J(Q^{\max}(X)) \to B^{\max}(X)$ is a homotopy equivalence if X is a p-torsion suspension of finite type and so it is a homotopy equivalence for any p-torsion suspension X. The proof is completed.

Proof of Theorem 1.6. It suffices to show that

$$\exp\big(\pi_*(\Omega\Sigma X)\big) \leq \max\{\exp\big(\pi_*(A^{\min}(X^{(n)}))\big); 1 \leq n < \infty\}.$$

By Theorem 6.5, there is a natural coalgebra decomposition

$$T(V^u) \cong \bigotimes_{n=1}^{\infty} A^{\min}(V^u, L_n^{\max})$$

for any $\mathbf{k}$-module V^u. Thus there is a natural decomposition

$$\Omega\Sigma X \simeq \prod_{n=1}^{\infty} A^{\min}(X, L_n^{\max})$$

for any p-torsion suspension X.

By Proposition 6.10, $A^{\min}(V^u; L_n^{\max})$ is a natural coalgebra retract of $A^{\min}(L_n^{\max}(V^u))$. Thus $A^{\min}(X, L_n^{\max})$ is a natural retract of $A^{\min}(L_n^{\max}(X))$ for any p-torsion suspension X.

Notice that $L_n^{\max}(V^u)$ is a natural retract of $(V^u)^{\otimes n}$. Thus $L_n^{\max}(X)$ is a natural retract of $X^{(n)}$ and so $A^{\min}(L_n^{\max}(X))$ is a natural retract of $A^{\min}(X^{(n)})$. Therefore $A^{\min}(X, L_n^{\max})$ is a natural retract of $A^{\min}(X^{(n)})$. In particular,

$$\exp\big(\pi_*(A^{\min}(X; L_n^{\max}))\big) \leq \exp\big(\pi_*(A^{\min}(X^{(n)}))\big).$$

Notice that

$$\pi_*(\Omega\Sigma X) \cong \sum_{n=1}^{\infty} \pi_*(A^{\min}(X, L_n^{\max}))$$

The assertion follows. □

10. The Functor L'_n and the Associated $\mathbf{k}(\Sigma_n)$-Module $\mathrm{Lie}'(n)$

In this section, we give a second natural coalgebra decomposition of the tensor algebra $T(V)$ (Theorem 10.7). The first factor of this decomposition will be $A^{\min}(V)$ as before, but the other factors will be tensor algebras. In contrast to the decomposition into minimal pieces obtained earlier, (Theorem 6.5), this decomposition collects factors into lumps which are convenient for certain purposes. This leads to the construction of a canonical projective $\mathbf{k}(S_n)$-submodule of $\mathrm{Lie}^{\max}(n)$ which we denote $\mathrm{Lie}'(n)$.

First, we recall some terminology from Section 4.

Let $M(V)$ be a natural submodule of $T(V)$. We write $A^{\min}(V; M)$ for the minimal coalgebra retract of $T(V)$ over $M(V)$. In other words, $A^{\min}(V; M)$ has the following properties:

(1). $M(V) \subseteq A^{\min}(V; M) \subseteq T(V)$.

(2). $A^{\min}(V;M)$ is a natural subcoalgebra of $T(V)$ with a coalgebra retraction $T(V)\to A^{\min}(V;M)$.
(3). Let $C(V)$ be any natural PROPER subcoalgebra of $A^{\min}(V;M)$ such that $M(V)\subseteq C(V)$. Then $C(V)$ is NOT a natural coalgebra retract of $T(V)$.

The existence of $A^{\min}(V;M)$ was proved in Section 4. The minimal coalgebra retract of $T(V)$ over V is denoted by $A^{\min}(V)$, i.e, $A^{\min}(V)=A^{\min}(V;V)$. Notice that $A^{\min}(V;M)$ is different from $A^{\min}(M)$ in general.

As in Section 6 we set $M^n(V)=\bigoplus_{j=1}^n L_j^{res}(V)\subseteq T(V)$. We write $A^{\min;n}(V)$ as an abbreviation for $A^{\min}(V;M^n)$. One can easily check that $A^{\min;n}(V)$ equals $A^{\min}(V;N^n)$ where $N^n(V)=\bigoplus_{j=1}^n V^{\otimes j}\subseteq T(V)$. In other words, $A^{\min;n}(V)$ is the natural minimal coalgebra retract of $T(V)$ which contains $T_j(V)$ for $j\le n$.

Let $r\colon T(V)\to A^{\min;n}(V)$ be a natural coalgebra retraction and let $s\colon A^{\min;n}(V)$ be the inclusion. One can define a new map

$$r'\colon T(V)\to A^{\min;n}(V)$$

by setting

(1). $r'\colon T_0(V)\to A_0^{\min;n}(V)$ is the identity map of $\mathbf{k}$.
(2).
$$r'(x_1\cdots x_m)=r(s(r'(x_1\cdots x_{m-1}))x_m)$$
for $m\ge 1$ and $x_j\in V$.

By Proposition 6.1, we have

(1). $r'\colon T(V)\to A^{\min;n}(V)$ is a natural map of coalgebras.
(2). $s\circ r'\colon T_j(V)\to T_j(V)$ is the identity for $j\le n$.
(3). $r'\circ s\colon A^{\min;n}(V)\to A^{\min;n}(V)$ is an isomorphism of coalgebras and so there is a coalgebra cross-section $s'\colon A^{\min;n}(V)\to T(V)$ such that $r'\circ s'$ is the identity map of $A^{\min;n}(V)$.
(4). there is a right $T(V)$-module structure on $A^{\min;n}(V)$ such that $r'\colon T(V)\to A^{\min;n}(V)$ is a morphism of right $T(V)$-modules.
(5). the kernel of $r'\colon T(V)\to A^{\min;n}(V)$ is a right ideal.
(6). the cotensor product
$$\mathbf{k}\square_{A^{\min;n}(V)}T(V)$$
under $r'\colon T(V)\to A^{\min;n}(V)$ is a sub-Hopf algebra of $T(V)$.

Notice that a natural coalgebra cross-section $s'\colon A^{\min;n}(V)\to T(V)$ of r' need NOT be morphism of right $T(V)$-modules. The natural coalgebra retraction $r'\colon T(V)\to A^{\min;n}(V)$ is stable in the following sense.

Proposition 10.1. *Let $r\colon T(V)\to A^{\min;n}(V)$ be a natural coalgebra retraction with a natural coalgebra cross-section $s\colon A^{\min;n}(V)\to T(V)$. Let $s'\colon A^{\min;n}(V)\to T(V)$ be a* *natural coalgebra crooss-section of* $r'\colon T(V)\to A^{\min;n}(V)$. *Let* r'' *denote* $(r')'\colon T(V)\to A^{\min;n}(V)$. *Then*

$$r''=r'\colon T(V)\to A^{\min;n}(V).$$

Proof. The proof is to show by induction on m that

$$r''=r'\colon T_j(V)\to A_j^{\min;n}(V)$$

for $j\leq m$. Notice that $r''=r'=r\colon T_j(V)\to A_j^{\min;n}(V)$ for $j\leq 1$. Suppose that $r''=r'\colon T_j(V)\to A_j^{\min;n}(V)$ for $j\leq m-1$ with $m>1$.

Let $x_1,\cdots,x_m\in V$. By definition,

$$r''(x_1\cdots x_m)=r'(s'(r''(x_1\cdots x_{m-1}))x_m).$$

Notice that

$$r'(\alpha x)=r(s(r'(\alpha))x)$$

for $\alpha\in T(V)$ and $x\in V$. Thus

$$\begin{aligned} r''(x_1\cdots x_m)&=r(s(r'(s'(r''(x_1\cdots x_{m-1}))))x_m)\\ &=r(s(r''(x_1\cdots x_{m-1}))x_m)\\ &=r(s(r'(x_1\cdots x_{m-1}))x_m)\\ &=r'(x_1\cdots x_m).\end{aligned}$$

The assertion follows. □

Lemma 6.3 shows that there is a natural coalgebra decomposition

$$A^{\min;n}(V)\cong A^{\min}(V;L_n^{\max}(V))\otimes A^{\min;n-1}(V)$$

for $n>1$. This produces a natural coalgebra decomposition of $T(V)$, see Theorem 6.5.

Lemma 10.2. *For each* $n\geq 1$, *there exists a natural coalgebra retraction*

$$p_n\colon T(V)\to A^{\min;n}(V)$$

such that

(1). $p_n\colon T(V)\to A^{\min;n}(V)$ *is a natural morphism of right* $T(V)$*-modules.*
(2). *there exists a natural morphism of coalgebra*

$$q_n\colon A^{\min;n}(V)\to A^{\min;n-1}(V)$$

such that the diagram

$$\begin{array}{ccc} T(V) & \xrightarrow{p_n} & A^{\min;n}(V) \\ \| & & \downarrow q_n \\ T(V) & \xrightarrow{p_{n-1}} & A^{\min;n-1}(V) \end{array}$$

commutes.

Proof. The proof is given by constructing natural coalgebra retractions

$$p_n\colon T(V) \to A^{\min;n}(V)$$

inductively. Let $p_1\colon T(V) \to A^{\min}(V)$ be a natural coalgebra retraction such that $p_1 = r'$ for some natural coalgebra retraction $r\colon T(V) \to A^{\min}(V)$. Suppose that the natural coalgebra retractions

$$p_j\colon T(V) \to A^{\min;j}(V)$$

have been defined, for $j < n$ with $n > 1$, such that

(1). $p_j = {r'}_j\colon T(V) \to A^{\min;j}(V)$ for some natural coalgebra retraction $r_j\colon T(V) \to A^{\min;j}(V)$.

(2). there is a natural morphism of coalgebras $q_j\colon A^{\min;j}(V) \to A^{\min;j-1}(V)$ such that

$$p_{j-1} = q_j \circ p_j\colon T(V) \to A^{\min;j-1}(V).$$

Let $B^{\max}(V; M^{n-1})$, or its abbreviation $B^{\max;n-1}(V)$, denote the cotensor product $\mathbf{k}\square_{A^{\min;n-1}(V)} T(V)$ using the comodule structure induced from $p_{n-1}\colon T(V) \to A^{\min;n-1}(V)$, and let $j\colon B^{\max;n-1}(V) \to T(V)$ be the natural inclusion. Then $B^{\max;n-1}(V)$ is sub Hopf algebra of $T(V)$. Notice that by the remarks preceding Lemma 6.2, $B^{\max;n-1}(V)$ is $(n-1)$-connected, with a natural module isomorphism $L_n^{\max}(V) \cong B_n^{\max;n-1}(V)$. Notice that $B^{\max;n-1}(V)$ is a natural coalgebra retract of $T(V)$ **over** $B_n^{\max;n-1}(V) = L_n^{\max}(V)$. Thus $A^{\min}(V; L_n^{\max})$ is a natural coalgebra retract of $B^{\max;n-1}(V)$ **over** $B_n^{\max;n-1}(V)$. Let $i\colon A^{\min}(V; L_n^{\max}) \to B^{\max;n-1}(V)$ be the natural inclusion and let $g\colon B^{\max;n-1}(V) \to A^{\min}(V; L_n^{\max})$ be a natural coalgebra retraction. Let $h\colon T(V) \to A^{\min}(V; L_n^{\max})$ be the composite

$$T(V) \xrightarrow{\phi} B^{\max;n-1}(V) \xrightarrow{g} A^{\min}(V; L_n^{\max}),$$

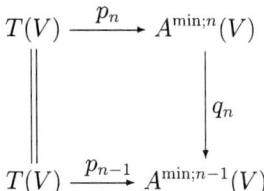

where $\phi\colon T(V) \to B^{\max;n-1}(V)$ is a natural coalgebra retraction. There is a commutative diagram

$$\begin{array}{ccccc}
T(V) & \xrightarrow{(h\otimes p_{n-1})\circ\psi} & A^{\min}(V; L_n^{\max}) \otimes A^{\min;n-1}(V) & \xrightarrow{\mu\circ((j\circ i)\otimes \bar{s}_{n-1})} & T(V) \\
\downarrow{\scriptstyle p_{n-1}} & & \downarrow{\scriptstyle \epsilon\otimes id_{A^{\min;n-1}(V)}} & & \downarrow{\scriptstyle p_{n-1}} \\
A^{\min;n-1}(V) & = & A^{\min;n-1}(V) & = & A^{\min;n-1}(V),
\end{array}$$

where $\bar{s}_{n-1}\colon A^{\min;n-1}(V) \to T(V)$ is a coalgebra cross-section of the retraction $p_{n-1}\colon T(V) \to A^{\min;n-1}(V)$. The right-side diagram commutes because

(1). the map $p_{n-1}\colon T(V) \to A^{\min;n-1}(V)$ is a morphism of right $T(V)$-modules.

(2). the injection $j\circ i\colon A^{\min}(V; L^{\max}) \to T(V)$ maps into $B^{\max;n-1}(V) = \mathbf{k}\square_{A^{\min;n-1}(V)} T(V)$.

Now let $f\colon T(V) \to T(V)$ denote the composite

$$T(V) \xrightarrow{(h\otimes p_{n-1})\circ\psi} A^{\min}(V; L_n^{\max}) \otimes A^{\min;n-1}(V) \xrightarrow{\mu\circ((j\circ i)\otimes \bar{s}_{n-1})} T(V).$$

Then $p_{n-1}\circ f = p_{n-1}\colon T(V) \to A^{\min;n-1}(V)$ and so the map $p_{n-1}\colon T(V) \to A^{\min;n-1}(V)$ factors through $\operatorname{colim}_f T(V)$. Notice that

$$\operatorname{colim}_f T(V) \cong A^{\min}(V; L^{\max}) \otimes A^{\min;n-1}(V) \cong A^{\min;n}(V)$$

as coalgebras. Thus there are natural coalgebra retractions $r_n\colon T(V) \to A^{\min;n}(V)$, $q_n\colon A^{\min;n}(V) \to A^{\min;n-1}(V)$ such that $p_{n-1} = q_n \circ r_n\colon T(V) \to A^{\min;n-1}(V)$. Let $s_n\colon A^{\min;n}(V) \to T(V)$ be a natural coalgebra cross-section of $r_n\colon T(V) \to A^{\min;n}(V)$. Then $p_{n-1}\circ s_n = q_n \circ r_n \circ s_n = q_n$. Let $p_n\colon T(V) \to A^{\min;n}(V)$ be defined by $p_n = r_n'\colon T(V) \to A^{\min;n}(V)$. We now show by a subsidiary induction that

$$q_n \circ p_n = p_{n-1}\colon T(V) \to A^{\min;n-1}(V).$$

Notice that $q_n\circ p_n = p_{n-1}\colon T_j(V) \to A_j^{\min;n-1}(V)$ for $j \leq 1$. Suppose that

$$q_n \circ p_n = p_{n-1}\colon T_j(V) \to A_j^{\min;n-1}(V)$$

for $j < m$ with $m > 1$. Let $x_i \in V$ with $1 \leq i \leq m$. Then

$$\begin{aligned}
q_n \circ p_n(x_1\cdots x_m) &= q_n \circ r_n'(x_1\cdots x_m) \\
&= q_n \circ r_n(s_n(r_n'(x_1\cdots x_{m-1}))x_m) \\
&= p_{n-1}(s_n(r_n'(x_1\cdots x_{m-1}))x_m) \\
&= r_{n-1}(s_{n-1}(p_{n-1}(s_n(r_n'(x_1\cdots x_{m-1}))))x_m) \\
&= r_{n-1}(s_{n-1}(q_n(r_n'(x_1\cdots x_{m-1})))x_m)
\end{aligned}$$

$$= r_{n-1}(s_{n-1}(p_{n-1}(x_1 \cdots x_{m-1}))x_m) = p_{n-1}(x_1 \cdots x_m),$$

where $s_{n-1}\colon A^{\min;n-1}(V) \to T(V)$ is a natural coalgebra cross-section of the coalgebra retraction $r_{n-1}\colon T(V) \to A^{\min;n-1}(V)$ and where $p_{n-1} = r'_{n-1}\colon T(V) \to A^{\min;n-1}$. Both inductions are now finished and the assertion follows. □

Proposition 10.3. (1). *For each $n \geq 1$, $B^{\max;n}(V)$ is a natural coalgebra-split subHopf algebra of $T(V)$. That is, $B^{\max;n}(V)$ is a natural subHopf algebra of $T(V)$ with a natural coalgebra retraction $T(V) \to B^{\max;n}(V)$.*

(2). *$B^{\max;n+1}(V)$ is natural coalgebra-split sub Hopf algebra of $B^{max;n}(V)$ for each $n \geq 1$ and so there is a descending chain of natural coalgebra-splitting sub Hopf algebras of $T(V)$*

$$\cdots \subseteq B^{\max;n}(V) \subseteq \cdots \subseteq B^{\max;1}(V) \subseteq T(V).$$

(3). *$B^{\max;n}(V)$ is an n-connected Hopf algebra with*

$$B^{\max;n}_{n+1}(V) \cong L^{\max}_{n+1}(V)$$

as modules

(4). *there is a natural isomorphism of coalgebras*

$$\mathbf{k} \otimes_{B^{\max;n+1}(V)} B^{\max;n}(V) \cong A^{\min}(V; L^{\max}_{n+1}).$$

(5). *there is a natural isomorphism of coalgebras*

$$B^{\max;n}(V) \cong \otimes_{j=n+1}^{\infty} A^{\min}(V; L^{\max}_j).$$

(6). *$B^{\max;n}(V)$ is a sub Hopf algebra of $T(V)$ generated by*

$$B^{\max;j}_{j+1}(V) \cong L^{\max}_{j+1}(V)$$

for $j \geq n$.

Proof. Assertions (1)-(5) follow immediately from Lemma 10.2. We need to show Assertion (6).

Let $\bar{B}$ denote the subHopf algebra of $T(V)$ generated by

$$B^{\max;j}_{j+1}(V) \cong L^{\max}_{j+1}(V)$$

for $j \geq n$. Then one has

$$\bar{B} \subseteq B^{\max;n}(V)$$

because $B^{\max;n}(V)$ is a sub Hopf algebra of $T(V)$ and

$$B^{\max;j}_{j+1}(V) \subseteq B^{\max;n}(V)$$

for $j \geq n$. Notice that $B^{\max;j}(V) \cap \bar{B}$ is a subHopf algebra of $T(V)$. There exists a unique morphism of algebras

$$\phi_j\colon T(B^{\max;j}_{j+1}(V)) \to B^{\max;j}(V) \cap \bar{B},$$

for each $j \geq n$, such that the restricted map

$$\phi_j|_{B^{\max;j}_{j+1}(V)} \colon B^{\max;j}_{j+1}(V) \to B^{max;j}(V) \cap \bar{B}$$

is the natural inclusion. Notice that $B^{\max;j}_{j+1}(V) \subseteq L_{j+1}(V)$. Thus

$$\phi_j \colon T(B^{\max;j}_{j+1}(V)) \to B^{\max;j}(V) \cap \bar{B},$$

is a morphism of Hopf algebras.

Let $H_{j+1} \colon T(V) \to T(V^{\otimes j+1})$ be the j-th James-Hopf map. Then the composite

$$T(B^{\max;j}_{j+1}(V)) \xrightarrow{\phi_j} B^{\max;j}(V) \cap \bar{B} \hookrightarrow T(V) \xrightarrow{H_{j+1}} T(V^{\otimes j+1})$$

is a morphism of Hopf algebras for $j \geq n$, [[17], Theorem 1.1]. Let

$$f_j \colon V^{\otimes j+1} \to B^{\max;j}_{j+1}(V)$$

be a natural retraction of **k**-modules. Then

$$T(f_j) \colon T(V^{\otimes j+1}) \to T(B^{\max;j}_{j+1}(V))$$

is a morphism of Hopf algebras and the composite

$$T(B^{\max;j}_{j+1}(V)) \xrightarrow{\phi_j} B^{\max;j}(V) \cap \bar{B} \hookrightarrow T(V) \xrightarrow{H_{j+1}} T(V^{\otimes j+1}) \xrightarrow{T(f_j)} T(B^{\max;j}_{j+1}(V))$$

is an isomorphism of Hopf algebras. Thus $T(B^{\max;j}_{j+1}(V))$ is a natural coalgebra retract of $B^{\max;j}(V) \cap \bar{B}$ for each $j \geq n$.

Notice that $A^{\min}(V; L^{\max}_{j+1})$ is a natural coalgebra retract of $T(B^{\max;j}_{j+1}(V))$, where $L^{\max}_{j+1}(V)$ is chosen to be $B^{\max;j}_{j+1}(V)$. Thus, for each $j \geq n$, there is a natural coalgebra injection

$$s_j \colon A^{\min}(V; L^{\max}_{j+1}) \to B^{\max;j}(V) \cap \bar{B}$$

with a coalgebra retraction

$$r_j \colon T(V) \to A^{\min}(V; L^{\max}_{j+1})$$

such that the composite

$$A^{\min}(V; L^{\max}_{j+1}) \xrightarrow{s_j} B^{\max;j}(V) \cap \bar{B} \hookrightarrow T(V) \xrightarrow{r_j} A^{\min}(V; L^{\max}_{j+1})$$

is the identity map of $A^{\min}(V; L^{\max}_{j+1})$. Let

$$\tilde{r}_j \colon T(V) \to A^{\min}(V; L^{\max}_{j+1})$$

be the composite

$$T(V) \xrightarrow{p_{j+1}} A^{\min;j+1}(V) \hookrightarrow T(V) \xrightarrow{r_j} A^{\min}(V; L^{\max}_{j+1})$$

for each $j \geq n$. Then one has

(1). the composite

$$A^{\min}(V;L^{\max}_{j+1})\overset{s_j}{\hookrightarrow}B^{\max;j}(V)\cap\bar B\hookrightarrow T(V)\xrightarrow{\tilde r_j}A^{\min}(V;L^{\max}_{j+1})$$

is a natural isomorphism of coalgebras **over** $L^{\max}_{j+1}=B^{\max;j}_{j+1}(V)$ for each $j\geq n$

(2). the composite

$$A^{\min}(V;L^{\max}_{i+1})\overset{s_i}{\hookrightarrow}B^{\max;i}(V)\cap\bar B\hookrightarrow T(V)\xrightarrow{\tilde r_j}A^{\min}(V;L^{\max}_{j+1})$$

is the trivial morphism of coalgebras for $i>j\geq n$.

Let

$$g\colon \otimes_{j=n}^{\infty}A^{\min}(V;L^{\max}_{j+1})\to\bar B$$

be the composite

$$\otimes_{j=n}^{\infty}A^{\min}(V;L^{\max}_{j+1})\xrightarrow{\otimes_{j=n}^{\infty}s_j}\otimes_{j=n}^{\infty}B^{\max;j}(V)\cap\bar B\xrightarrow{\mu}\bar B,$$

where the multiplication map

$$\mu\colon \otimes_{j=n}^{\infty}B^{\max;j}(V)\cap\bar B\to\bar B$$

is well-defined because the connectivity of $B^{\max;j}(V)$ is convergent to ∞. Let

$$h\colon T(V)\to\otimes_{j=n}^{\infty}A^{\min}(V;L^{\max}_{j+1})$$

be the composite

$$T(V)\xrightarrow{\psi}\otimes_{j=n}^{\infty}T(V)\xrightarrow{\otimes_{j+n}^{\infty}\tilde r_j}\otimes_{j=n}^{\infty}A^{\min}(V;L^{\max}_{j+1}).$$

Let

$$\theta\colon \otimes_{j=n}^{\infty}A^{\min}(V;L^{\max}_{j+1})\to\otimes_{j=n}^{\infty}A^{\min}(V;L^{\max}_{j+1})$$

be the composite

$$\otimes_{j=n}^{\infty}A^{\min}(V;L^{\max}_{j+1})\xrightarrow{g}\bar B\hookrightarrow T(V)\xrightarrow{h}\otimes_{j=n}^{\infty}A^{\min}(V;L^{\max}_{j+1}).$$

Notice that the composite

$$A^{\min}(V;L^{\max}_{i+1})\hookrightarrow\otimes_{j=n}^{\infty}A^{\min}(V;L^{\max}_{j+1})\xrightarrow{\theta}\otimes_{j=n}^{\infty}A^{\min}(V;L^{\max}_{j+1})\xrightarrow{proj.}A^{\min}(V;L^{\max}_{j+1})$$

is an isomorphism for $i=j\geq n$ and trivial for $i>j\geq n$. Thus

$$P(\theta)\colon \oplus_{j=n}^{\infty}P(A^{\min}(V;L^{\max}_{j+1}))\to\oplus_{j=n}^{\infty}P(A^{\min}(V;L^{\max}_{j+1}))$$

is an isomorphism and so

$$\theta\colon \otimes_{j=n}^{\infty}A^{\min}(V;L^{\max}_{j+1})\to\otimes_{j=n}^{\infty}A^{\min}(V;L^{\max}_{j+1})$$

is a monomorphism. Thus the natural map θ is an isomorphism for V of finite type and so is a natural isomorphism for any V. Therefore

$$\otimes_{j=n}^{\infty} A^{\min}(V; L_{j+1}^{\max})$$

is a natural coalgebra retract of $\bar{B}$.

Notice that there is a natural isomorphism of coalgebras

$$\otimes_{j=n}^{\infty} A^{\min}(V; L_{j+1}^{\max}) \cong B^{\max;n}(V).$$

If V is of finite type, then the Poincaré series satisfy

$$\chi(B^{\max;n}(V)) \geq \quad \chi(\bar{B}) \quad \geq \chi(\otimes_{j=n}^{\infty} A^{\min}(V; L_{j+1}^{\max})) = \chi(B^{\max;n}(V)).$$

Thus

$$\bar{B} = B^{\max;n}(V)$$

if V is of finite type and so $\bar{B} = B^{\max;n}(V)$ for any V. The assertion follows. □

Let L be a Lie algebra. We write $U(L)$ for the universal enveloping algebra of L. The following lemma is well known. (See [5].)

Lemma 10.4. (1). *Let L' be a sub Lie algebra of a Lie algebra L. Then $U(L')$ is a sub Hopf algebra of $U(L)$.*

(2). *If L' is a Lie ideal of a Lie algebra L, then $U(L')$ is a normal sub Hopf algebra of $U(L)$.*

(3). *Let*

$$0 \longrightarrow L' \longrightarrow L \longrightarrow L'' \longrightarrow 0$$

be a short exact sequence of augmented Lie algebras. Then

$$U(L') \hookrightarrow U(L) \twoheadrightarrow U(L'')$$

is a short exact sequence of Hopf-algebras.

Let $n \geq 1$ and let $LB^{\max;n}(V)$ denote $B^{\max;n}(V) \cap L(V)$. Then $LB^{\max;n}(V)$ is a natural sub Lie algebra of $L(V)$.

Proposition 10.5. *There is a natural isomorphism of Hopf algebras*

$$U(LB^{\max;n}(V)) \cong B^{\max;n}(V)$$

for each $n \geq 1$.

Proof. The inclusion

$$j\colon LB^{\max;n}(V) \to B^{\max;n}(V)$$

induces an unique morphism of Hopf algebras

$$\phi\colon U(LB^{\max;n}(V)) \to B^{\max;n}(V)$$

such that

$$\phi|_{LB^{\max;n}(V)} = j\colon LB^{\max;n}(V) \to B^{\max;n}(V).$$

There is a commutative diagram of Hopf algebras

$$\begin{array}{ccc} B^{\max;n}(V) & \hookrightarrow & T(V) \\ \uparrow \phi & & \uparrow \cong \\ U(LB^{\max;n}(V)) & \hookrightarrow & U(L(V)). \end{array}$$

Thus the map

$$\phi\colon U(LB^{\max;n}(V)) \to B^{\max;n}(V)$$

is a monomorphism. By assertion (6) of Proposition 10.3, the map

$$\phi\colon U(LB^{\max;n}) \to B^{\max;n}(V)$$

is an epimorphism. The assertion follows. □

Lemma 10.6 ([17], Theorem1.1). *Let $H_n\colon T(V) \to T(V^{\otimes n})$ be the n-th James-Hopf invariant. Let $C(n)$ denote the sub Hopf algebra of $T(V)$ generated by $L_j(V)$ for $j \geq n$. Then the composite*

$$C(n) \hookrightarrow T(V) \xrightarrow{H_n} T(V^{\otimes n})$$

is a morphism of Hopf algebras.

Theorem 10.7. *For each $n \geq 1$, there exists a natural submodule $L_j^{\prime;n}(V)$ of $L_j(V)$ for $j \geq n+1$ such that there is a natural isomorphism of coalgebras*

$$T(V) \cong \otimes_{j=1}^{n} A^{\min}(V; L_j^{\max}(V) \otimes \otimes_{j=n+1}^{\infty} T(L_j^{\prime;n}(V)).$$

Proof. It suffices to show that there exists a natural submodule $L_j^{\prime;n}(V)$ of $L_j(V)$ for $j \geq n+1$ such that there is a natural isomorphism of coalgebras

$$B^{\max;n}(V) \cong \otimes_{j=n+1}^{\infty} T(L^{\prime;n}(V)).$$

This statement follows from the following lemma. □

Lemma 10.8 (Semi-tensor Product Resolution of $B^{\max;n}(V)$). *There exists a descending chain $\{B^{(j;n)}(V)\}_{j\geq n+1}$ of sub Hopf algebra of $T(V)$ such that*

(1). *$B^{(n+1;n)}(V) = B^{\max;n}(V)$.*
(2). *$B^{(j;n)}(V)$ is $(j-1)$-connected for each $j \geq n+1$.*
(3). *$B^{(j;n)}(V)$ is a natural coalgebra retract of $T(V)$ for each $j \geq n+1$.*

(4). *There is a natural isomorphism of Hopf algebras*

$$U(LB^{(j;n)}(V)) \cong B^{(j;n)}(V)$$

where $LB^{(j;n)}(V) = B^{(j;n)}(V) \cap L(V)$.

(5). *Let*

$$L'^{;n}_j(V) = B^{(j;n)}_j(V)$$

for $j \geq n+1$. *Then there is a natural short exact sequence of Hopf algebras*

$$B^{(j+1;n)}(V) \hookrightarrow B^{(j;n)}(V) \xrightarrow{\phi_n} T(L'^{;n}_j(V)).$$

Furthermore the morphism of Hopf algebras

$$\phi_n \colon B^{(j;n)} \to T(L'^{;n}_j(V))$$

has a **Hopf algebra** *cross-section*

$$T(L'^{;n}_j(V)) \hookrightarrow B^{(j;n)}(V).$$

Proof. The proof will be given by constructing $B^{(j;n)}(V)$ inductively on j for $j \geq n+1$ starting with

$$B^{(n+1;n)}(V) = B^{\max;n}(V).$$

By Proposition 10.5, we have

$$U(LB^{(n+1;n)}(V)) \cong B^{(n+1;n)}(V).$$

Let $m \geq n+1$. Suppose that $B^{(j;n)}(V)$ have been defined for $n+1 \leq j \leq m$ with the properties stated in the Lemma. We need to construct $B^{(m+1;n)}(V)$.

Notice that $B^{(m;n)}(V)$ is $(m-1)$-connected and $B^{(m;n)}(V)$ is a natural coalgebra retract of $T(V)$. Thus

(1). $L'^{;n}_m(V) = B^{(m;n)}_m(V)$ is a natural submodule of $L_m(V)$.

(2). $L'^{;n}_m(V)$ is a natural module retract of $V^{\otimes m}$.

Let

$$\phi_m \colon B^{(m;n)}(V) \to T(L'^{;n}_m(V)$$

be the composite

$$B^{(m;n)}(V) \hookrightarrow T(V) \xrightarrow{H_m} T(V^{\otimes m}) \xrightarrow{T(r)} T(L'^{;n}_m(V)),$$

where $r \colon V^{\otimes m} \to L'^{;n}_m(V)$ is a natural retraction of modules. Notice that

$$U(LB^{(m;n)}(V)) \cong B^{(m;n)}(V)$$

by induction. Thus one has

$$B^{(m;n)}(V) \subseteq C(m),$$

where $C(m)$ is the sub Hopf algebra of $T(V)$ generated by $L_j(V)$ for $j \geq m$. By Lemma 10.6, the composite

$$C(m) \hookrightarrow T(V) \xrightarrow{H_m} T(V^{\otimes m})$$

is a morphism of Hopf algebras. Thus the composite

$$B^{(m;n)}(V) \hookrightarrow C(m) \hookrightarrow T(V) \xrightarrow{H_m} T(V^{\otimes m})$$

is a morphism of Hopf algebras and so the natural map

$$\phi_m\colon B^{(m;n)}(V) \to T(L_m^{\prime;n}(V))$$

is a morphism of Hopf algebras.

Let

$$f_m\colon T(L_m^{\prime;n}(V)) \to B^{(m;n)}(V)$$

be the morphism of Hopf algebras induced by the inclusion

$$L_m^{\prime;n}(V) = B_m^{(m;n)}(V) \subseteq B^{(m;n)}(V).$$

Then the composite $\phi_m \circ f_m$, which is a morphism of Hopf algebras, is the identity map of $T(L_m^{\prime;n}(V))$.

Let $B^{(m+1,n)}(V)$ be defined by the cotensor product

$$B^{(m+1;n)}(V) = \mathbf{k}\square_{T(L_m^{\prime;n}(V))} B^{(m;n)}(V)$$

using the comodule structure induced from the map $p_m\colon B^{(m;n)}(V) \to T(L_m^{\prime;n}(V)$. Then one gets a natural short exact sequence of Hopf algebras

$$B^{(m+1;n)}(V) \hookrightarrow B^{(m;n)}(V) \overset{\phi_m}{\twoheadrightarrow} T(L_m^{\prime;n}(V)),$$

with a **Hopf algebra** cross-section

$$f_m\colon T(L_m^{\prime;n}(V)) \to B^{(m;n)}(V).$$

Notice that

$$\begin{aligned} LB^{(m+1;n)}(V) =& B^{(m+1;n)}(V) \cap L(V) \\ =& (\operatorname{Ker}\phi_m\colon B^{(m;n)}(V) \to T(L_m^{\prime;n}(V)) \cap L(V) \\ =& \operatorname{Ker}\phi_m\colon LB^{(m;n)}(V) \to T(L_m^{\prime;n}(V)). \end{aligned}$$

We have

$$L(V) \cap T(L_m^{\prime;n}(V)) = L(L_m^{\prime;n}(V))$$

because

(1). $L(L_m^{\prime;n}(V)) \subseteq L(V) \cap T(L_m^{\prime;n}(V))$.
(2). $U(L_m^{\prime;n}(V)) = U(L(V) \cap T(L_m^{\prime;n}(V)) = T(L_m^{\prime;n}(V))$.

(3). Using the Poincaré-Birkhoff-Witt Theorem, we see that the Poincaré Series satisfy

$$\chi(L(L'^{;n}_m(V))) = \chi(L(V) \cap T(L'^{;n}_m(V))).$$

when V is finite dimensional.

Let L'' be the image

$$L'' = \operatorname{Im}\phi_m \colon LB^{(m;n)}(V) \to T(L'^{;n}_m(V).$$

Then

$$L(L'^{;n}_m(V)) \subseteq L''$$

because

$$f_m(L(L'^{;n}_m(V))) \subseteq LB^{(m;n)}(V)$$

and the composite $\phi_m \circ f_m$ is the identity map of $T(L'^{;n}_m(V))$.

Let $\theta\colon T(V) \to T(V)$ be the composite

$$T(V) \xrightarrow{H_m} T(V^{\otimes m}) \xrightarrow{T(r)} T(L'^{;n}_m(V)) \hookrightarrow T(V).$$

We show that

$$\theta(L(V)) \subseteq L(V).$$

Let $x_1, \cdots, x_q \in V$. Let $\bar{V}$ be a connected graded module generated by the graded words $x_1, \cdots x_q$ and let $\tilde{\gamma}_q \subseteq \bar{V}^{\otimes q}$ be a graded module generated by graded monomials

$$x_{\sigma(1)} \cdots x_{\sigma(q)}$$

for $\sigma \in \Sigma_q$. Then

$$\theta([[x_1, \cdots, x_q]) \in \tilde{\gamma}_q \cap L^{res}_q(\bar{V})) \subseteq L_q(\bar{V}).$$

Thus $\theta(L(V)) \subseteq L(V)$ for any V. Notice that $LB^{(m;n)}(V) \subseteq L(V)$. Thus

$$\phi_m(LB^{(m;n)}(V)) \subseteq L(V) \cap T(L'^{;n}_m(V)) = L(L'^{;n}_m(V))$$

and so

$$L'' \subseteq L(L'^{;n}_m(V)).$$

We have a short exact sequence of Lie algebras

$$0 \longrightarrow LB^{(m+1;n)}(V) \to LB^{(m;n)}(V) \to L(L'^{;n}_m(V)) \longrightarrow 0,$$

and thus there is a short exact sequence of Hopf algebras

$$U(LB^{(m+1;n)}(V)) \hookrightarrow U(LB^{(m;n)}(V)) \twoheadrightarrow U(L(L'^{;n}_m(V))).$$

Consider the commutative diagram of short exact sequences of Hopf algebras

$$\begin{array}{ccccc}
B^{(m+1;n)}(V) & \hookrightarrow & B^{(m;n)}(V) & \twoheadrightarrow & T(L'^{;n}_m(V)) \\
\uparrow & & \uparrow\cong & & \uparrow\cong \\
U(LB^{(m+1;n)}(V)) & \hookrightarrow & U(LB^{(m;n)}(V)) & \twoheadrightarrow & U(L(L'^{;n}_m(V))).
\end{array}$$

This yields the isomorphism of Hopf algebras

$$U(LB^{(m+1;n)}(V)) \cong B^{(m+1;n)}(V).$$

The induction is finished now and the assertion follows. □

11. Examples

11.1. The functor $A^{\min}_n$ for $n \leq p$.

Let V be a connected graded $\mathbf{k}$-module. Let $\mathrm{Sym}(V)$ be the free commutative (graded) algebra generated by V and let $\mathrm{Sym}_n(V)$ be the $\mathbf{k}$-submodule of $\mathrm{Sym}(V)$ generated by the homogeneous elements of degree n.

Proposition 11.1. *Let V be a connected graded $\mathbf{k}$-module. Then*

1) *$A^{\min}_n(V)$ is isomorphic to $\mathrm{Sym}_n(V)$ for $n < p$.*
2) *$A^{\min}_p(V)$ is isomorphic to $\mathrm{Sym}_{p-1}(V) \otimes V$*

Proof. If $n < p$, by Proposition 6.1 and Theorem 8.4, we have

$$A^{\min}_n(V^u) \cong \mathbf{k} \otimes_{\mathbf{k}(S_n)} (V^u)^{(\otimes n)}$$

for any ungraded $\mathbf{k}$-module V^u. Thus

$$A^{\min}_n(V) \cong \mathbf{k} \otimes_{\mathbf{k}(S_n)} V^{\otimes n}$$

for $n < p$ and any connected graded $\mathbf{k}$-module V, where S_n acts on $V^{\otimes n}$ by permuting factors in **graded** sense. Assertion 1) follows.

By Corollary 8.27, $A^{\min}_p(V^u) \cong \mathbf{k} \otimes_{\mathbf{k}(S_{p-1})} (V^u)^{(\otimes p)}$ for any ungraded $\mathbf{k}$-module V^u. Thus

$$A^{\min}_p(V) \cong \mathbf{k} \otimes_{\mathbf{k}(S_{p-1})} V^{\otimes p}$$

for any connected graded $\mathbf{k}$-module V, where S_{p-1} acts on $V^{\otimes p}$ by permuting factors in **graded** sense. Assertion 2) follows. □

Recall that the fuctor $A^{\min}_n$ is geometrically realizable when $\mathbf{k} = \mathbb{Z}/p\mathbb{Z}$. (See Section 9.) If $n \leq p$, then the space $A^{\min}_n(X)$ can be constructed using a transfer map. Suppose $n \leq m$ and let S_n act on $X^{(m)}$ by permuting the first n factors. Let X be a suspension and let $\phi_n \colon X^{(m)} \longrightarrow X^{(m)}$ be a map whose homotopy class is given by $[\phi_n] = \sum_{\sigma \in S_n}[\sigma]$ in the group $[X^{(m)}, X^{(m)}]$, where $\sigma\colon X^{(m)} \longrightarrow X^{(m)}$ for $\sigma \in S_n$ is the action map. One can use a fixed choice of order for the addition of the $n!$ maps to define the map ϕ_n. The map ϕ_n defined in this way will be **strictly** functorial in Y for $X = \Sigma Y$.

Corollary 11.2. *Let X be a p-torsion suspension. Then*

1) $A^{\min}_n(X) \simeq \operatorname{hocolim}_{\phi_n} X^{(n)}$ *for $n < p$;*
2) $A^{\min}_p(X) \simeq \operatorname{hocolim}_{\phi_{p-1}} X^{(p)}$.

Let $Sp^n(X)$ be the n-fold symmetric product of X.

Proposition 11.3. *Let X be a p-torsion suspension. Then*

1) *$A^{\min}_n(X)$ is homotopy equivalent to $Sp^n(X)/Sp^{n-1}(X)$ for $n < p$;*
2) *$A^{\min}_p(X)$ is homotopy equivalent to $(Sp^{p-1}(X)/Sp^{p-2}(X)) \wedge X$.*

Proof. Let $q\colon X^{(n)} \to Sp^n(X)/Sp^{n-1}(X)$ be the quotient map. Then the map q factors through $A^{\min}_n(X)$. The assertion follows by checking mod p and rational homology. □

11.2. The functor $B^{\max}$.

Recall that $B^{\max}(V)$ is a sub Hopf algebra of $T(V)$. By Theorem 1.1, $L_n(V) \subseteq B^{\max}(V)$ if n is not a power of p. Thus the sub Hopf algebra of $T(V)$ generated by $L_n(V)$ for $n \neq p^t$ for some $t \geq 0$, which is denoted by $B(V)$, is contained in $B^{\max}(V)$. Now we show by examples that $B(V)$ is a proper sub Hopf algebra of $B^{\max}(V)$.. Notice that, by Corollary 8.9, $B^{\max}(V)$ is isomorphic to $T(\oplus_{n=2}^{\infty} Q^{\max}_n(V))$. By Proposition 8.10 and Corollary 8.27, $Q^{\max}_p(V) = 0$. Thus $B(V)$ is isomorphic to $B^{\max}(V)$ up to tensor weight $p^2 - 1$. We will show that $Q^{\max}_9(V) \neq 0$ when $p = 3$, $Q^{\max}_4(V) = 0$ when $p = 2$ and $Q^{\max}_8(V) \neq 0$ when $p = 2$. This show that $B^{\max}(V)$ differs from $B(V)$ in tensor length of p^2 when $p = 3$ and in tensor length 8 when $p = 2$.

First we consider the case where $p = 2$. Notice that $\mathrm{Lie}(4)$ is of dimension 6. Recall that any projective S_n module is a free $\mathrm{Syl}_p(S_n)$-module, where $\mathrm{Syl}_p(S_n)$ is a Sylow p-subgroup of S_n. In particular, the order of $\mathrm{Syl}_p(S_n)$ divides the dimension of a projective S_n-module. It follows that $\mathrm{Lie}^{\max}(4) = 0$, that is, $\mathrm{Lie}(4)$ does not have a non-trivial projective S_4-submodule. Thus $Q^{\max}_4(V) \subseteq L^{\max}_4(V) = 0$. Let W be the two dimensional $\mathbf{k}$-module generated by u, v. Let $L^{(i)}_n(W)$ be the $\mathbf{k}$-submodule of $L_n(W)$ generated by the Lie elements $[[x_1, x_2], \cdots, x_n]$, where $x_i = u$ or v, and u occurs i times in the sequence $(x_1, \cdots, x_n)$. Let $r_V\colon T(V) \to A^{\min}(V)$ be the functorial coalgebra retraction.

Proposition 11.4. *Suppose the ground field* $\mathbf{k}$*is of characteristic* 2. *Then the image of $L^{(2)}_8(W)$ in $A^{\min}(W)$ under the retraction r_W is a one dimensional* $\mathbf{k}$*-module.*

Proof. Notice that $\dim L_8^{(2)}(W) = 3$. Let $\phi\colon V^{\otimes 8} \longrightarrow V^{\otimes 8}$ be the composite

$$V^{\otimes 8} \xrightarrow{T_{2,5}} V^{\otimes 8} \xrightarrow{\beta_8} V^{\otimes 8},$$

where $T_{2,5}$ is the map which switches the second position and the fifth position and $\beta_n(x_1 \otimes \cdots \otimes x_n) = [[x_1, x_2, \cdots, x_n]$. Then ϕ is a functorial map and so $\operatorname{colim}_\phi V^{\otimes 8}$ is a functorial retract of $V^{\otimes 8}$. Notice that the map ϕ factors through $L_8(V)$. Thus $\operatorname{colim}_\phi V^{\otimes 8}$ is a functorial retract of $L_8^{\max}(V)$.

By direct calculation, we have $\dim \operatorname{colim}_\phi W^{\otimes 8} \geq 2$. It follows that

$$\dim \operatorname{Im}(L_8^{(2)}(W) \to A_8^{\min}(W)) \leq 1.$$

On the other hand, let $\theta\colon T(V) \longrightarrow T(V)$ be **any** functorial map of coalgebras such that $\theta|_V$ is the identity. Then θ is geometrically realizable. Let $V = \bar{H}_*(P^n(2))$ be the homology of mod 2 Moore space. Then

$$\theta([[v, u, u, u, u, u, u, u]) = [[v, u, u, u, u, u, u, u]$$

by Lemma 8.19. Thus $\theta([[v, u, v, u, u, u, u, u]) \neq 0$ because

$$Sq^1_*([[v, u, v, u, u, u, u, u]) = [[v, u, u, u, u, u, u, u].$$

It follows that $\theta([[u, v, u, v, v, v, v, v]) \neq 0$. Thus

$$\operatorname{Im}(L_8^{(2)}(W) \to A_8^{\min}(W)) \neq 0$$

and the assertion follows. □

Proposition 11.5. *Suppose the ground field* $\mathbf{k}$ *is of characteristic* 2. *Then* $Q_8^{\max}(V) \neq 0$ *if* $\dim V \geq 2$ *and so* $B(V)$ *is a proper sub Hopf algebra of* $B^{\max}(V)$ *if* $\dim V \geq 2$.

Proof. The functors $Q_n^{\max}(V)$ for $n < 8$ are as follows:

$Q_2^{\max}(V) = Q_4^{\max}(V) = 0$, $Q_n^{\max}(V) = L_n(V)$ for $n = 3, 5, 7$ and $Q_6^{\max}(V) \cong L_6(V)/[L_3(V), L_3(V)]$.

Let $L'(V)$ be the sub Lie algebra of $L(V)$ generated by $Q_n^{\max}$ for $n < 8$. Then $L'(W) \cap L_8^{(2)}(W)$ is of dimension one, generated by $[[[u, v, v], [u, v, v, v, v]]$. By Proposition 11.4, $\dim Q_8^{\max}(W) \cap L_8^{(2)}(W) = 1$. The assertion follows. □

The case when $p = 3$ and $n = 9$ is similar. We let W be the two dimensional $\mathbf{k}$ with basis u and v and let $Z \subset W^{\otimes 9}$ be the intersection of $\mathrm{L}_9(W)$ with the submodule of tensors with two u's and seven v's. Let $\phi : V^{\otimes 9} \to V^{\otimes 9}$ be the composite $\beta_9 \circ T_{2,7}$, where $T_{2,7}$ denotes the map which interchanges positions 2 and 7. Then $\dim Z = 4$ and direct calculation shows that $\dim \operatorname{colim}_\phi = 3$. But if $B^{\max}(W)$ were equal to $B(W)$ then $\dim \operatorname{colim}_\phi$ could be at most 2.

11.3. **The symmetric group module** $\mathrm{Lie}^{\max}(p)$.

Recall that $\mathrm{Lie}^{\max}(n)$ is the maximal projective S_n-submodule of $\mathrm{Lie}(n)$. Now we determine $\mathrm{Lie}^{\max}(p)$.

Let $\bar{L}(V)$ be the sub Lie algebra of the free Lie algebra $L(V)$ generated by $L_n(V)$ for $1 < n < p$. Let $\bar{V}$ be p-dimensional $\mathbf{k}$-module generated by the $\{x_1, \cdots, x_p\}$. Notice that $\mathrm{Lie}(p) \subseteq \gamma_p \subseteq \bar{V}^{\otimes p}$.

Proposition 11.6. *There is an isomorphism of S_p-modules*

$$L^{\max}(p) \cong \bar{L}(\bar{V}) \cap \mathrm{Lie}(p).$$

Proof. By Theorem 6.5, there is a functorial isomorphism

$$T(V) \cong \bigotimes_{n=1}^{\infty} A^{\min}(V; L_n^{\max}).$$

It follows that there is a functorial isomorphism

$$L_p(V) \cong \bigoplus_{n=1}^{\infty} A^{\min}(V; L_n^{\max}) \cap L_p(V) = \bigoplus_{n=1}^{p} A^{\min}(V; L_n^{\max}) \cap L_p(V).$$

Notice that $A^{\min}(V; L_n^{\max})$ is a functorial retract of $T(L_n^{\max}(V))$. Thus $A_k^{\min}(V; L_n^{\max}) = 0$ when k is not divisible by n. Hence $A^{\min}(V; L_n^{\max}) \cap L_p(V) = 0$ when $1 < n < p$ and therefore $L_p(V) \cong (A^{\min}(V) \cap L_p(V)) \oplus L_p^{\max}(V)$. It follows that there is a functorial isomorphism

$$L_p^{\max}(V) \cong B^{\max}(V) \cap L_p(V).$$

The assertion follows. □

Example 11.7. Let $p = 5$. There is a functorial isomorphism

$$L_5^{\max}(V) \cong [L_2(V), L_3(V)].$$

Thus $\mathrm{Lie}^{\max}(5)$ is isomorphic to the sub $\mathbf{k}$-module of $\mathrm{Lie}(5)$ generated by

$$[[x_{i_1}, x_{i_2}], [[x_{i_3}, x_{i_4}], x_{i_5}]]$$

for $1 \leq i_s \leq 5$.

Example 11.8. Let $p = 7$. Then $L_7^{\max}(V)$ is isomorphic to the submodule

$$[L_2(V), L_5(V)] + [L_3(V), L_4(V)]$$

of $L_7(V)$.

Corollary 11.9. *The* $\mathbf{k}$*-module* $\mathrm{Lie}^{\max}(p)$ *is of dimension* $(p-1)! - p + 1$.

11.4. Calculations for small n when $p = 2$.

We have shown that in general we have inclusions of $\mathbf{k}(S_n)$-modules $0 \subset \mathrm{Lie}'(n) \subset \mathrm{Lie}^{\max}(n) \subset \mathrm{Lie}(n)$, where the first three are projective $\mathbf{k}(S_n)$-modules, with $\mathrm{Lie}^{\max}(n)$ the maximum projective $\mathbf{k}(S_n)$-module. In this section we give some calculations of $\mathrm{Lie}'(n)$ and $\mathrm{Lie}^{\max}(n)$ for small values of n, when $p = 2$.

Recall that the dimension of a projective $\mathbf{k}(G)$-module must be divisible by the order of a Sylow p-subgroup of G. (See end of proof of Proposition 8.29.)

Now let $p = 2$. If n is odd, then β_n is an idempotent (in characteristic 2) and so $\mathrm{Lie}(n)$ itself is projective in this case. If $n = 2$ or $n = 4$, then the order of the Sylow 2-subgroup of S_n is greater than the dimension of $\mathrm{Lie}(n)$ and so $\mathrm{Lie}(n)$ has no proper projective submodules in these cases. Therefore we have

$$0 \subsetneq \mathrm{Lie}'(1) = \mathrm{Lie}^{\max}(1) = \mathrm{Lie}(1),$$

$$0 = \mathrm{Lie}'(2) = \mathrm{Lie}^{\max}(2) \subsetneq \mathrm{Lie}(2),$$

$$0 \subsetneq \mathrm{Lie}'(3) = \mathrm{Lie}^{\max}(3) = \mathrm{Lie}(3),$$

$$0 = \mathrm{Lie}'(4) = \mathrm{Lie}^{\max}(4) \subsetneq \mathrm{Lie}(4), \qquad \text{and}$$

$$0 \subsetneq \mathrm{Lie}'(5) = \mathrm{Lie}^{\max}(5) = \mathrm{Lie}(5).$$

The first interesting case is $n = 6$. As we shall see, all of the containments are strict in this case. Notice that the order of the Sylow 2-subgroup of S_6 is 16, which does not divide $\dim \mathrm{Lie}(6) = 120$. Therefore the last containment must be strict.

By Theorem 10.7, there is a functorial decomposition

$$T(V) \cong A^{\min}(V) \otimes \bigotimes_{n=2}^{\infty} T(L'_n(V)),$$

where $L'_j = L'^{;1}_j$. By Theorem 8.3, $A^{\min}(V)$ does not have primitive elements of tensor length 6. It follows that

$$L_6(V) \cong [L_3(V), L_3(V)] \oplus L'_6(V) \qquad \text{or} \qquad L'_6(V) \cong L_6(V)/[L_3(V), L_3(V)].$$

Notice that $L'_n(V)$ is a functorial summand of $L^{\max}_n(V)$. Therefore $L_6(V)/[L_3(V), L_3(V)]$ is a functorial summand of $L^{\max}_6(V)$. Applying this to the 6-dimensional $\mathbf{k}$-module $\bar{V}$ with basis $\{x_1, x_2, \cdots, x_6\}$, and intersecting with γ_6 gives $\mathrm{Lie}'(6) = \mathrm{Lie}(6)/[L_3, L_3]$, where $[L_3, L_3]$ is an abbreviation for $[L_3(\bar{V}), L_3(\bar{V})] \cap \gamma_6$. It is easy to check that $\dim[L_3, L_3] = 40$, and so $\dim \mathrm{Lie}'(6) = 80$. We note that it does not seem to be easy to give a direct proof (i.e. without using the methods of this paper) that the $\mathbf{k}(S_6)$-module $\mathrm{Lie}(6)/[L_3, L_3]$ is projective.

Next we consider $\mathrm{Lie}^{\max}(6)$. From the preceding, we know that this is the same as computing the largest projective $\mathbf{k}(S_6)$-submodule of $[L_3, L_3]$. Let $\phi\colon V^{\otimes 6} \to V^{\otimes 6}$ be the functorial map defined by

$$\phi(a_1a_2\cdots a_6) = [[[a_1, a_4], a_3], [[a_2, a_5], a_6]]$$

for $a_j \in V$. Then $\mathrm{colim}_\phi \bar{V}^{\otimes 6} \cap \gamma_6$ is a non-trivial projective submodule of $[L_3(\bar{V}), L_3(\bar{V})] \cap \gamma_6$. (See Example 6.11.)

Thus we have

Proposition 11.10. *Let* $\mathbf{k}$ *be of characteristic* 2. *Then*

$$L_6(V)/[L_3(V), L_3(V)] \oplus \mathrm{colim}_\phi V^{\otimes 6}$$

is a functorial summand of $L_6^{\max}(V)$, *and the second term is nontrivial (while the first has dimension* 80*).*

Corollary 11.11. *The* $\mathbf{k}$*-module* $\mathrm{Lie}^{\max}(6)$ *is of dimension* 96 *or* 112.

Remark 11.12. *By using known facts about the representations of* S_6 *and further calculations, we can show that there is no* S_6*-projective submodule of* $\mathrm{Lie}(6)$ *which is of dimension* 112. *Thus* $\dim(\mathrm{Lie}^{\max}(6)) = 96$ *and there is a functorial isomorphism*

$$L_6^{\max}(V) \cong L_6(V)/[L_3(V), L_3(V)] \oplus \mathrm{colim}_\phi V^{\otimes 6}.$$

Continuing,

$$0 \subsetneq \mathrm{Lie}'(7) = \mathrm{Lie}^{\max}(7) = \mathrm{Lie}(7).$$

Since $\mathrm{Lie}'(2) = \mathrm{Lie}'(4) = 0$, for $n = 8$ we get

$$0 \subsetneq \mathrm{Lie}'(8) = \mathrm{Lie}^{\max}(8) \subsetneq \mathrm{Lie}(8).$$

We have not succeeded in calculating $\mathrm{Lie}^{\max}(8)$ although after much calculation we have narrowed it down to two possibilities, having dimensions 3840 and 4224 respectively.

11.5. Decompositions of $\Omega\Sigma^2 X$ for two-cell complexes X.

While $A^{\min}(X)$ is the smallest **natural** retract of $\Omega\Sigma X$, for an **individual** X it is possibly that it could decompose further. We will now explore this situation for the case of the mod p Moore space.

Example 11.13. Suppose $p > 2$ and let $X = P^n(p^r)$ be the n-dimensional mod p^r Moore space. Let Z be the smallest retract of the space $\Omega\Sigma X$ which contains the bottom cell. Then Z is a retract of $A^{\min}(X)$. Z was computed by Cohen, Moore, and Neisendorfer in [6],[14]. We will compare Z with $A^{\min}(X)$.

If n is odd checking homology shows $A^{\min}(X)$ is homotopy equivalent to Z up to dimension $p(n-1)$ and $A^{\min}(X)$ first differs from Z in dimension $p(n-1)+1$.

Let F be the homotopy fibre of the retraction $A^{\min}(X) \to Z$. Then F is homotopy equivalent to $P^{p(n-1)+2}(p^r)$ up to dimension $p^2(n-1)$.

If n is even checking homology shows $A^{\min}(X)$ is homotopy equivalent Z up to dimension $p^2(n-1)$. Thus the first possible difference in homotopy between $A^{\min}(X)$ and Z is in dimension $p^2(n-1)+1$. In this case, we show that $A^{\min}(X)$ differs from Z by finding a primitive element of dimension $(p-2)(pn-2)+2(p-1)$ in $H_*(A^{\min}(X))$ which is not in $H_*(Z)$.

Let V be the graded $\mathbb{Z}/p$-module generated by two elements u and v, where u and v are of dimensions $n-1$ and n, respectively. Let W be the graded $\mathbb{Z}/p$-module generated by u and v', where v' is of dimension $n-2$. Let $f\colon V \to W$ be the **ungraded** map of $\mathbb{Z}/p$-modules defined by setting $f(u) = u$ and $f(v) = v'$. Notice that $f^{\otimes n}\colon V^{\otimes n} \longrightarrow W^{\otimes n}$ is an isomorphism of S_n-modules for each n, where S_n acts on $V^{\otimes n}$ and $W^{\otimes n}$ by permuting factors in graded sense. Thus f induces an isomorphism

$$f = A^{\min}(f)\colon A^{\min}(V) \xrightarrow{\cong} A^{\min}(W).$$

Now we assume that n is even. Following the notation in [5] we set $\tau_k(v) = \mathrm{ad}^{p^k-1}(v)(u)$, and

$$\sigma_k(v) = (1/2)\sum_{j=1}^{p^k-1} p^{-1}\binom{p^k}{j}[\mathrm{ad}^{j-1}(v)(u), \mathrm{ad}^{p^k-j-1}(v)(u)],$$

Let ϕ_V be the composite $T(V) \xrightarrow{r_V} A^{\min}(V) \hookrightarrow T(V)$, where r_V is a functorial coalgebra retraction. By Lemma 8.19, $\phi_V(\tau_k) = \tau_k$ for each $k \geq 1$. Let $\tilde{V}$ be a free graded $\mathbb{Z}$-module generated by $\tilde{u}$ and $\tilde{v}$ with $|\tilde{v}| = n$ and $|\tilde{u}| = n-1$. Then there is a functorial coalgebra map $\tilde{\phi}\colon T(\tilde{V}) \to T(\tilde{V})$ such that $\tilde{\phi} \otimes \mathbb{Z}/p = \phi\colon T(\tilde{V}) \otimes_{\mathbb{Z}} \mathbb{Z}/p = T(V) \to T(\tilde{V}) \otimes_{\mathbb{Z}} \mathbb{Z}/p = T(V)$. Since $\tau_k(\tilde{v})$ is a fixed point of $\tilde{\phi} \mod (p)$, it follows that $\tilde{\phi}(\tau_k(\tilde{v})) = a_k\tau_k(\tilde{v})$ for each $k \geq 1$, where $a_k \equiv 1 \mod (p)$, Let $d\colon T(\tilde{V}) \to T(\tilde{V})$ be the (graded) derivation such that $d(\tilde{v}) = \tilde{u}$. Then $\tilde{\phi} \circ d = d \circ \tilde{\phi}$. Notice that $d(\tau_k(\tilde{v})) = p \cdot \sigma_k(\tilde{v})$ in $T(\tilde{V})$. (See [5].) Thus $\tilde{\phi}(\sigma_k(\tilde{v})) = a_k\sigma_k(\tilde{v})$. It follows that $\sigma_k(v)$ is a fixed point of $\phi_V\colon T(V) \to T(V)$. Thus

$$\sigma_k(v') = f(\sigma_k(v)) = (1/2)\sum_{j=1}^{p^k-1} p^{-1}\binom{p^k}{j}[\mathrm{ad}^{j-1}(v')(u), \mathrm{ad}^{p^k-j-1}(v')(u)]$$

is a fixed point of $\phi_W\colon T(W) \to T(W)$. Now $T(W) = H_*(\Omega\Sigma P^{n-1}(p^r))$. Let β^r be the r-th Bockstein. Then

$\beta^r(\sigma_k(v')) = (1/2)\sum_{j=1}^{p^k-1} p^{-1}\binom{p^k}{j}[\beta^r(\mathrm{ad}^{j-1}(v')(u)), \mathrm{ad}^{p^k-j-1}(v')(u)]$
$+(-1)^{|\mathrm{ad}^{j-1}(v')(u)|}[\mathrm{ad}^{j-1}(v')(u), \beta^r(\mathrm{ad}^{p^k-j-1}(v')(u))].$

Notice that $|u|$ is odd, $|v'|$ is even and $\beta^r(u) = v'$. Thus

$$\beta^r(\sigma_k(v')) = (1/2)\left(p^{k-1}[v', \mathrm{ad}^{p^k-2}(v')(u)] - p^{k-1}[\mathrm{ad}^{p^k-2}(v')(u), v']\right)$$
$$= p^{k-1}[v', \mathrm{ad}^{p^k-2}(v')(u)] = p^{k-1}\,\mathrm{ad}^{p^k-1}(v')(u),$$

where we used the fact that $ad^j(v')(v') = 0$ for $j \geq 1$. In particular, $\beta^r(\sigma_1(v') = \mathrm{ad}^{p-1}(v')(u)$. Notice that $\mathrm{ad}^{p-1}(v')(u)$ is spherical. Thus there exists a map $g\colon P^m(p^r) \to \Omega\Sigma P^{n-1}(p^r)$ such that $g_*(v_m) = \sigma_1(v')$, where v_m is a generator for $H_m(P^m(p^r)) = \mathbb{Z}/p$ and $m = |\sigma_1(v')| = (p-2)(n-2)+2(n-1)$. Let $\tilde{g}\colon \Omega\Sigma P^m(p^r) \longrightarrow \Omega\Sigma P^{n-1}(p^r)$ be the map of H-spaces such that $\tilde{g}|_{P^m(p^r)} = g$. By [6, Lemma 4.7], the map $\tilde{g}$ has a left homotopy inverse $r\colon \Omega\Sigma P^{n-1}(p^r) \longrightarrow \Omega\Sigma P^m(p^r)$. Let θ be the composite

$$\Omega\Sigma P^m(p^r) \xrightarrow{\tilde{g}} \Omega\Sigma P^{n-1}(p^r) \xrightarrow{\phi} \Omega\Sigma P^{n-1}(p^r) \xrightarrow{r} \Omega\Sigma P^m(p^r),$$

where $\phi\colon \Omega\Sigma P^{n-1}(p^r) \to \Omega\Sigma P^{n-1}(p^r)$ is a geometric realization of the functorial coalgebra map $\phi_W\colon T(W) \to T(W)$. Then $\theta_*\colon H_m(\Omega\Sigma P^m(p^r)) \longrightarrow H_m(\Omega\Sigma P^m(p^r))$ is an isomorphism. Let Z' be the smallest retract of $\Omega\Sigma P^m(p^r)$. Notice that the map $\phi\colon \Omega\Sigma P^{n-1}(p^r) \to \Omega\Sigma P^{n-1}(p^r)$ factors through $A^{\min}(P^{n-1}(p^r))$. Thus Z' is a retract of $A^{\min}(P^{n-1}(p^r))$. Notice that m is an even integer. Thus $H_*(Z')$, as a coalgebra, is isomorphic to the (graded) free commutative algebra generated by u_{m-1}, v_m, $\tau_l(v_m)$ and $\sigma_l(v_m)$ for $l \geq 1$. Let

$$\tau_l(\sigma_1(v')) = \mathrm{ad}^{p^l-1}(\sigma_1(v'))(\mathrm{ad}^{p-1}(v')(u));$$

$$\sigma_l(\sigma_1(v')) = (1/2)\sum_{j=1}^{p^l-1} p^{-1}\binom{p^l}{j}[\mathrm{ad}^{j-1}(\sigma_1(v'))(\mathrm{ad}^{p-1}(v')(u)), \mathrm{ad}^{p^l-j-1}$$
$$(\sigma_1(v'))(\mathrm{ad}^{p-1}(v')(u))].$$

Then $\phi_W(\tau_l(\sigma_1(v'))$ and $\phi_W(\sigma_l(\sigma_1(v')))$ are nonzero for $l \geq 1$. Thus the elements

$$\tau_l(\sigma_1(v)) = f^{-1}(\tau_l(\sigma_1(v')))) = \mathrm{ad}^{p^l-1}(\sigma_1(v))(\mathrm{ad}^{p-1}(v)(u)) \qquad \text{and}$$

$$\sigma_l(\sigma_1(v)) = (1/2)\sum_{j=1}^{p^l-1} p^{-1}\binom{p^l}{j}[\mathrm{ad}^{j-1}(\sigma_1(v))(\mathrm{ad}^{p-1}(v)(u)), \mathrm{ad}^{p^l-j-1}(\sigma_1(v))$$
$$(\mathrm{ad}^{p-1}(v)(u))]$$

for $l \geq 1$ have non-trivial images in $A^{\min}(V)$ under the coalgebra retraction $r_V\colon T(V) \to A^{\min}(V)$. Notice that $H_*(Z)$, as a coalgebra, is isomorphic to the free commutative algebra generated by $u, v, \tau_k(v), \sigma_k(v)$ for $k \geq 1$. It follows that the elements $\tau_l(\sigma_1(v))$ and $\sigma_l(\sigma_1(v)$ have trivial image in $H_*(Z)$ under any coalgebra retraction $T(V) \to H_*(Z)$. In particular, the element $\sigma_1(\sigma_1(v))$, which is of dimension $(p-2)(pn-2)+$

$2(pn-1)$, induces a primitive element in $H_*(A^{\min}(X))$ which has a trivial image in $H_*(Z)$ under any coalgebra retraction $H_*(A^{\min}(X)) \to H_*(Z)$.

Next we consider 2-cell complexes in which the cells are not in adjacent degrees.

Let p be an **odd** prime and let X be a p-local two-cell complex. Let u, v be a basis for $\bar{H}_*(X)$ with $|v| > |u|$. We consider four cases. First we consider the case where both $|u|$ and $|v|$ are odd. The following example was provided by Fred Cohen.

Proposition 11.14. *Let X be finite p-local suspension such that 1) $\bar{H}_{\text{even}}(X) = 0$ and 2) $\dim(\bar{H}_*(X)) < p-1$. That is, the number of cells in X is less than p. Then $H_*(A^{\min}(X))$ is isomorphic as a coalgebra to the exterior algebra generated by $\bar{H}_*(X)$.*

The proof is immediate by checking the Poincaré series.

Corollary 11.15. *Suppose $p > 3$ and let X be a p-local two-cell suspension. Let u, v be a basis for $\bar{H}_*(X)$ with $|v| > |u|$. If both $|v|$ and $|u|$ are odd, then $B^{\max}(X)$ is homotopy equivalent to $\Omega\Sigma(L_2(X) \vee L_3(X))$ and $H_*(A^{\min}(X))$ is isomorphic, as a coalgebra, to the exterior algebra generated by u, v.*

The proof is immediate by checking the Poincaré series.

Corollary 11.16. *Let p be an odd prime and let X be finite p-local suspension such that 1) $\bar{H}_{\text{even}}(X) = 0$ and 2) $\dim(\bar{H}_*(X)) < p-1$. Then the p-torsion component of $\pi_*(A^{\min}(X))$ has a bounded exponent.*

Notice that the p-torsion component of $\pi_*(\Omega\Sigma X)$ does NOT have a bounded exponent in general. This example shows that the smallest retract $A^{\min}(X)$ does have a bounded exponent.

The case where both $|u|$ and $|v|$ are even is more complicated. Partial information is given in Proposition 11.1, Corollary 11.2 and Proposition 11.3. Some special properties are as follows. Let $q\colon \Sigma X \to S^{|v|+1}$ be the pinch map and let F_q be the homotopy fibre of q.

Proposition 11.17. *Let $p > 3$ and let X be a p-local two-cell suspension. Let u, v be a basis for $\bar{H}_*(X)$ with $|v| > |u|$. Assume that both $|u|$ and $|v|$ are even. Then*

1) *$S^{|u|+1}$ is a retract of F_q.*
2) *There exists a $p-1$-cell complex Y such that there is a fibre sequence*

$$Y \longrightarrow A^{\min}(X) \longrightarrow \Omega(A^{\min}(\Sigma X))$$

   up to dimension $p^2|u|$.
3) *ΩF_q is homotopy equivalent $\Omega S^{|u|+1} \times \Omega\Sigma(\vee_{n=2}^{\infty} Q_n^{\max}(X) \vee Y)$ up to dimension $p^2|u|$.*

4) *There exist a map* $\phi\colon L_p^{\max}(X)\vee Y\to \Omega\Sigma X$ *such that* $\bar{H}_*(L_p^{\max}(X)\vee Y)$ *is isomorphic to the submodule* $L_p(\bar{H}_*(X))$. *In other words,* $L_p(\bar{H}_*(X))$ *is geometrically realizable.*
5) *There is a fibre sequence*

$$\Omega\Sigma(\bigvee_{n=2}^{\infty} Q_n^{\max}(X)\vee Y)\longrightarrow \Omega\Sigma X\longrightarrow \Omega(A^{\min}(\Sigma X))$$

up to dimension $p^2|u|$.

Proof. Notice that $H_*(A^{\min}(\Sigma X))$ is the exterior algebra generated by $\bar{H}_*(\Sigma X)$. Thus there is a fibre sequence

$$A^{\min}(S^{|u|+1})=S^{|u|+1}\longrightarrow A^{\min}(\Sigma X)\longrightarrow A^{\min}(S^{|v|+1})=S^{|v|+1}.$$

Therefore there is a homotopy commutative diagram of fibre sequence

$$\begin{array}{ccccc}
S^{|u|+1} & \longrightarrow & A^{\min}(\Sigma X) & \longrightarrow & S^{|v|+1}\\
\uparrow & & \cup\!\!\uparrow & & \Vert\\
F_q & \longrightarrow & \Sigma X & \longrightarrow & S^{|v|+1}.
\end{array}$$

Assertion 1) follows.

Let Y be the $p|v|-1$ skeleton of the homotopy fibre of the composite

$$A^{\min}(X)\hookrightarrow \Omega\Sigma X\longrightarrow \Omega(A^{\min}(\Sigma X)).$$

Then assertions 2)-5) follow by checking homology. □

We now the consider the case where $|v|$ is odd and $|u|$ is even. (See Example 11.13 for discussion of the opposite parity.) In this case, there is a decomposition of $\Omega\Sigma X$ which can be regarded as a generalization of the classical decomposition of odd dimensional Moore spaces [5, 6].

Proposition 11.18. *Let* p *be an odd prime and let* X *be a* p*-local two-cell suspension. Let* u, v *be a basis for* $\bar{H}_*(X)$ *with* $|v|>|u|$. *If* $|v|$ *is odd and* $|u|$ *is even, then there is a homotopy equivalence*

$$\Omega\Sigma X\simeq \Omega\Sigma(\bigvee_{n=0}^{\infty}\Sigma^{n|u|+|v|}X)\times F$$

where the mod p *homology* $H_*(F)$ *is isomorphic, as a coalgebra, to the free commutative alegbra generated by* $\bar{H}_*(X)$.

Proof. When $|v| = |u| + 1$, then X is a wedge of two spheres or a Moore space. The answer is known in this case [5, 6]. Thus we assume that $|v| > |u| + 1$. Let $V = \bar{H}_*(X)$.

Notice that $L_2(V)$ has a basis $[u, v], v^2 = [v, v]/2$. Since $p > 2$, $L_2(X) = L_2^{\max}(X)$ is a retract of $X^{(2)}$. FRom the cofibre sequence

$$S^{|v|-1} \wedge X \longrightarrow S^{|u|} \wedge X \longrightarrow X \wedge X,$$

one gets $L_2(X) \simeq \Sigma^{|v|} X$. Let ϕ_n be the composite

$$S^{n|u|} \wedge L_2(X) \hookrightarrow X^{(n)} \wedge L_2(X) \hookrightarrow X^{(n+2)} \xrightarrow{W_n} \Omega\Sigma X,$$

where $S^{n|u|} \to X^{(n)}$ is the inclusion of the bottom cell and W_n is the n-fold Samelson product from *right to left.* Let

$$\phi \colon \bigvee_{n=0}^{\infty} S^{n|u|} \wedge L_2(X) \longrightarrow \Omega\Sigma X$$

be the map defined by $\phi|_{S^{n|u|} \wedge L_2(X)} = \phi_n$. Let

$$J(\phi) \colon \Omega\Sigma(\bigvee_{n=0}^{\infty} S^{n|u|} \wedge L_2(X)) \longrightarrow \Omega\Sigma X$$

be the map of H-spaces such that

$$J(\phi)|_{\bigvee_{n=0}^{\infty} S^{n|u|} \wedge L_2(X)} = \phi.$$

Now consider the homotopy commutative diagram of fibre sequences

$$\begin{array}{ccccccc}
\Omega S^{|u|+1} & \longrightarrow & F & \longrightarrow & S^{|v|} & \xrightarrow{\Sigma f} & S^{|u|+1} \\
\downarrow{\scriptstyle \Omega i} & & \downarrow{\scriptstyle \theta} & & \downarrow & & \downarrow{\scriptstyle i} \\
\Omega\Sigma X & = & \Omega\Sigma X & \longrightarrow & * & \longrightarrow & \Sigma X,
\end{array}$$

where $S^{|v|-1} \xrightarrow{f} S^{|u|} \xrightarrow{i} X$ is the cofibre sequence. It is routine to show that the map

$$\mu \circ (J(\phi) \times \theta) \colon \Omega\Sigma(\bigvee_{n=0}^{\infty} \Sigma^{n|u|} L_2(X)) \times F \longrightarrow \Omega\Sigma X$$

is a homotopy equivalence, where μ is the multiplication of $\Omega\Sigma X$. Notice that $H_*(F)$ is isomorphic, as a coalgebra, to the free commutative algebra generated by u, v. □

Corollary 11.19. *Let p be an odd prime and let X be a p-local two-cell suspension. Let u, v be a basis for $\bar{H}_*(X)$ with $|v| > |u|$. Suppose that $|v|$ is odd and $|u|$ is even. Let F be the smallest retract of the individual space $\Omega\Sigma X$ which contains the bottom cell. Then the p-torsion component of $\pi_*(F)$ has a bounded exponent.*

11.6. The PBW map in characteristic 0.

Let the ground field $\mathbf{k}$ be of characteristic 0. By Proposition 6.12, there is a functorial isomorphism of coalgebras $T(V) \cong \bigotimes_{n=1}^{\infty} S(L_n(V))$, where $S(W)$ is the free (graded) commutative algebra generated by W. We construct specific maps of coalgebras

$$\phi\colon T(V) \to \bigotimes_{n=1}^{\infty} S(L_n(V)) \text{ and } \theta\colon \bigotimes_{n=1}^{\infty} S(L_n(V)) \to T(V)$$

such that ϕ and θ are isomorphisms.

The map ϕ is defined by the composite

$$\begin{aligned} T(V) \xrightarrow{\quad\psi\quad} \bigotimes^{\infty} T(V) \xrightarrow{\otimes H_n} \bigotimes_{n=1}^{\infty} T(V^{\otimes n}) \\ \xrightarrow{\otimes T(\frac{1}{n}\beta_n)} \bigotimes_{n=1}^{\infty} T(L_n(V)) \xrightarrow{\otimes q_n} \bigotimes_{n=1}^{\infty} S(L_n(V)), \end{aligned}$$

where ψ is the comultiplication, H_n is the James-Hopf map and q_n is the quotient map.

Let W be a connected graded module. Let $\mathrm{tr}\colon S(W) \to T(W)$ be the transfer, more precisely

$$\mathrm{tr}(a_1 a_2 \cdots a_n) = \frac{1}{n!} \sum_{\sigma \in S_n} \sigma \cdot (a_1 a_2 \cdots a_n),$$

where S_n acts on $W^{\otimes n}$ in **graded** sense. It is easy to check that the transfer tr is a morphism of coalgebras. The map θ is defined by the composite

$$\bigotimes_{n=1}^{\infty} S(L_n(V)) \xrightarrow{\otimes \mathrm{tr}} \bigotimes_{n=1}^{\infty} T(L_n(V)) \hookrightarrow \bigotimes^{\infty} T(V) \xrightarrow{\mu} T(V),$$

where μ is the multiplication.

Proposition 11.20. *The maps ϕ and θ defined above are isomorphisms of coalgebras.*

Proof. We may assume that V is of finite type. Notice that $T(V)$ and $\bigotimes_{n=1}^{\infty} S(L_n(V))$ have same Poincaré series. Thus it suffices to show that the composite

$$\phi\circ\theta\colon \bigotimes_{n=1}^{\infty} S(L_n(V)) \longrightarrow \bigotimes_{n=1}^{\infty} S(L_n(V))$$

is a monomorphism. Notice that the module of primitive elements $P(\bigotimes_{n=1}^{\infty} S(L_n(V)))$ equals $\bigoplus_{n=1}^{\infty} L_n(V)$. Let $f = (\phi\circ\theta)|_{P(\bigotimes_{n=1}^{\infty} S(L_n(V)))}\colon \bigoplus_{n=1}^{\infty} L_n(V) \to \bigoplus_{n=1}^{\infty} L_n(V)$ and let $\pi_j\colon \bigoplus_{n=1}^{\infty} L_n(V) \to L_j(V)$ be the projection. Then we have

1) $\pi_n\circ f|_{L_n(V)}\colon L_n(V)\to L_n(V)$ is the identity map for each $n\geq 1$;
2) $\pi_j\circ f|_{L_n(V)}\colon L_n(V)\to L_j(V)$ is zero for any $1\leq n<j$.

Thus the map f is an isomorphism. The assertion follows. □

Remark 11.21. *The map θ is NOT the inverse of ϕ.*

The functorial map θ evaluated on $\bigotimes_{j=1}^{\infty} S(L_n(\bar{V}))$, where $\bar{V}$ is an n-dimensional $\mathbf{k}$-module generated by $\{x_1,\cdots,x_n\}$, induces the representations of S_n discussed in [16, Section 8.5]. Proposition 11.20 gives a much shorter proof of the canonical decomposition. In addition, by choosing V to be a graded module, Proposition 11.20 gives the canonical decomposition of "super" symmetric group modules, which are very useful in mathematical physics. Another interesting thing is that the James-Hopf map is used in defining the map ϕ.

In characteristic 0, a functorial isomorphism $R: T(V)\cong\bigotimes_{n=1}^{\infty} S(L_n(V))$ (although not a coalgebra isomorphism) has been previously given in [16] by Reutenauer who refers to it as the canonical decomposition. It is interesting to compare R with θ^{-1}. Each yields for each n an idempotent (different from β_n) whose image is Lie(n). For example, when $n=3$, R results in the identity

$$\begin{aligned} abc &= \big((a+b+c)^3-(a+b)^3-(a+c)^3-(b+c)^3+a^3+b^3+c^3)\big)/6\\ &\quad+\big((a+[b,c])^2-a^2-[b,c]^2\big)+(b+[a,c])^2-b^2-[a,c]^2\big)\\ &\qquad\quad+(c+[a,b])^2-c^2-[a,b]^2\big)/4\\ &\quad+[[a,b],c]/3-[[a,c],b]/6 \end{aligned}$$

which simplifies to

$$\begin{aligned} abc &= (abc+acb+bac+bca+cab+cba)/6\\ &\quad+(a[b,c]+b[a,c]+c[a,b]+[b,c]a+[a,c]b+[a,b]c)/4\\ &\quad+[[a,b],c]/3-[[a,c],b]/6 \end{aligned}$$

yielding the idempotent $abc \mapsto [[a,b],c]/3 - [[a,c],b]/6$. In comparison, θ^{-1} results in the identity

$$\begin{aligned} abc &= (abc + acb + bac + bca + cab + cba)/6 \\ &+ (a[b,c] + b[a,c] + c[a,b])/2 \\ &+ [[a,b],c]/3 + [[a,c],b]/3 \end{aligned}$$

yielding the idempotent $abc \mapsto [[a,b],c]/3 + [[a,c],b]/3$.

References

[1] H. Cartan and S. Eilenberg, *Homological algebras*, Princeton Univ. Press, Princeton, N.J., (1956).

[2] F. R. Cohen, *Applications of loop spaces to some problems in topology*, Lond. Math. Soc. Lect. Notes in Math, **139**, (1985) 11-20.

[3] F. Cohen, *On combinatorial group theory in homotopy*, Contemp. Math., **188** (1995), 57-63.

[4] F. Cohen, *On combinatorial groups in homotopy theory*, (to appear).

[5] F. R. Cohen, J. C. Moore and J. A. Neisendorfer, *Torsion in homotopy groups*, Ann. of Math., **109** (1979), 121-168

[6] F. R. Cohen, J. C. Moore and J. A. Neisendorfer, *The double suspension and exponents of homotopy groups of spheres*, Ann. of Math., **110** (1979), 549-565.

[7] F. R. Cohen, J. C. Moore and J. A. Neisendorfer, *Exponents in homotopy theory*, Algebraic topology and algebraic K-theory (Princeton, N.J., 1983), Ann. of Math. Stud., **113** (1987), 3-34.

[8] F. R. Cohen and L. R. Taylor, *Homology of function spaces*, Math. Z. **198** (1988), 299-319.

[9] F. R. Cohen and J. Wu, *A remark on the homotopy groups of* $\Sigma^n\mathbb{R}P^2$, Contem. Math. **181** (1995) 65-81

[10] G. Fröbenius, *Gesammelte Abhandlungen*, Springer, Berlin, (1968).

[11] P. J. Hilton and U. Stammbach, *A course in homological algebra*, Springer-Verlag

[12] J. Milnor and J. Moore, *On the structure of Hopf Algebras*, Ann. of Math. **81** (1965), 211-264.

[13] J. Moore and F. Peterson, *Nearly Fröbenius algebras, Poincaré algebras and their modules*, J. Pure. Appl. Algebra **3** (1973), 83-93.

[14] J. A. Neisendorfer, *3-primary exponents*, Proc. Camb. Phil. Soc., **90** (1981), 63-83.

[15] P. S. Selick, *Odd primary torsion in* $\pi_k S^3$, Topology, **17** (1978), 407-412.

[16] C. Reutenauer, *Free Lie Algebras*, Clarendon Press. Oxford, (1993)

[17] J. Wu, *On combinatorial calculations for the James-Hopf maps*, Topology, (to appear).

[18] J. Wu, *On products on minimal simplicial sets*, (to appear).

Department of Mathematics, University of Toronto, Toronto, Ontario M5G 3G3, Canada, selick@math.toronto.edu

Department of Mathematics, University of Pennsylvania, Philadelphia, PA 19104, USA, jiewu@math.upenn.edu

Editorial Information

To be published in the *Memoirs*, a paper must be correct, new, nontrivial, and significant. Further, it must be well written and of interest to a substantial number of mathematicians. Piecemeal results, such as an inconclusive step toward an unproved major theorem or a minor variation on a known result, are in general not acceptable for publication. Papers appearing in *Memoirs* are generally longer than those appearing in *Transactions*, which shares the same editorial committee.

As of July 31, 2000, the backlog for this journal was approximately 9 volumes. This estimate is the result of dividing the number of manuscripts for this journal in the Providence office that have not yet gone to the printer on the above date by the average number of monographs per volume over the previous twelve months, reduced by the number of volumes published in four months (the time necessary for preparing a volume for the printer). (There are 6 volumes per year, each containing at least 4 numbers.)

A Consent to Publish and Copyright Agreement is required before a paper will be published in the *Memoirs*. After a paper is accepted for publication, the Providence office will send a Consent to Publish and Copyright Agreement to all authors of the paper. By submitting a paper to the *Memoirs*, authors certify that the results have not been submitted to nor are they under consideration for publication by another journal, conference proceedings, or similar publication.

Information for Authors

Memoirs are printed from camera copy fully prepared by the author. This means that the finished book will look exactly like the copy submitted.

The paper must contain a *descriptive title* and an *abstract* that summarizes the article in language suitable for workers in the general field (algebra, analysis, etc.). The *descriptive title* should be short, but informative; useless or vague phrases such as "some remarks about" or "concerning" should be avoided. The *abstract* should be at least one complete sentence, and at most 300 words. Included with the footnotes to the paper should be the 2000 *Mathematics Subject Classification* representing the primary and secondary subjects of the article. The classifications are accessible from `www.ams.org/msc/`. The list of classifications is also available in print starting with the 1999 annual index of *Mathematical Reviews*. The Mathematics Subject Classification footnote may be followed by a list of *key words and phrases* describing the subject matter of the article and taken from it. Journal abbreviations used in bibliographies are listed in the latest *Mathematical Reviews* annual index. The series abbreviations are also accessible from `www.ams.org/publications/`. To help in preparing and verifying references, the AMS offers MR Lookup, a Reference Tool for Linking, at `www.ams.org/mrlookup/`. When the manuscript is submitted, authors should supply the editor with electronic addresses if available. These will be printed after the postal address at the end of the article.

Electronically prepared manuscripts. The AMS encourages electronically prepared manuscripts, with a strong preference for $\mathcal{AMS}$-LaTeX. To this end, the Society has prepared $\mathcal{AMS}$-LaTeX author packages for each AMS publication. Author packages include instructions for preparing electronic manuscripts, the *AMS Author Handbook*, samples, and a style file that generates the particular design specifications of that publication series. Though $\mathcal{AMS}$-LaTeX is the highly preferred format of TeX, author packages are also available in $\mathcal{AMS}$-TeX.

Authors may retrieve an author package from e-MATH starting from `www.ams.org/tex/` or via FTP to `ftp.ams.org` (login as `anonymous`, enter username as password, and type `cd pub/author-info`). The *AMS Author Handbook* and the *Instruction Manual* are available in PDF format following the author packages link from `www.ams.org/tex/`. The author package can be obtained free of charge by sending email to `pub@ams.org` (Internet) or from the Publication Division, American Mathematical Society, P.O. Box 6248, Providence, RI 02940-6248. When requesting an author package, please specify AMS-LaTeX or AMS-TeX, Macintosh or IBM (3.5) format, and the publication in which your paper will appear. Please be sure to include your complete mailing address.

Sending electronic files. After acceptance, the source file(s) should be sent to the Providence office (this includes any TeX source file, any graphics files, and the DVI or PostScript file).

Before sending the source file, be sure you have proofread your paper carefully. The files you send must be the EXACT files used to generate the proof copy that was accepted for publication. For all publications, authors are required to send a printed copy of their paper, which exactly matches the copy approved for publication, along with any graphics that will appear in the paper.

TeX files may be submitted by email, FTP, or on diskette. The DVI file(s) and PostScript files should be submitted only by FTP or on diskette unless they are encoded properly to submit through email. (DVI files are binary and PostScript files tend to be very large.)

Electronically prepared manuscripts can be sent via email to `pub-submit@ams.org` (Internet). The subject line of the message should include the publication code to identify it as a Memoir. TeX source files, DVI files, and PostScript files can be transferred over the Internet by FTP to the Internet node `e-math.ams.org` (130.44.1.100).

Electronic graphics. Comprehensive instructions on preparing graphics are available at `www.ams.org/jourhtml/graphics.html`. A few of the major requirements are given here.

Submit files for graphics as EPS (Encapsulated PostScript) files. This includes graphics originated via a graphics application as well as scanned photographs or other computer-generated images. If this is not possible, TIFF files are acceptable as long as they can be opened in Adobe Photoshop or Illustrator. No matter what method was used to produce the graphic, it is necessary to provide a paper copy to the AMS.

Authors using graphics packages for the creation of electronic art should also avoid the use of any lines thinner than 0.5 points in width. Many graphics packages allow the user to specify a "hairline" for a very thin line. Hairlines often look acceptable when proofed on a typical laser printer. However, when produced on a high-resolution laser imagesetter, hairlines become nearly invisible and will be lost entirely in the final printing process.

Screens should be set to values between 15% and 85%. Screens which fall outside of this range are too light or too dark to print correctly. Variations of screens within a graphic should be no less than 10%.

Inquiries. Any inquiries concerning a paper that has been accepted for publication should be sent directly to the Electronic Prepress Department, American Mathematical Society, P. O. Box 6248, Providence, RI 02940-6248.

Editors

This journal is designed particularly for long research papers (and groups of cognate papers) in pure and applied mathematics. Papers intended for publication in the *Memoirs* should be addressed to one of the following editors. In principle the Memoirs welcomes electronic submissions, and some of the editors, those whose names appear below with an asterisk (*), have indicated that they prefer them. However, editors reserve the right to request hard copies after papers have been submitted electronically. Authors are advised to make preliminary email inquiries to editors about whether they are likely to be able to handle submissions in a particular electronic form.

Algebra to CHARLES CURTIS, Department of Mathematics, University of Oregon, Eugene, OR 97403-1222 email: `cwc@darkwing.uoregon.edu`

Algebraic geometry and commutative algebra to LAWRENCE EIN, Department of Mathematics, University of Illinois, 851 S. Morgan (M/C 249), Chicago, IL 60607-7045; email: `ein@uic.edu`

Algebraic topology and cohomology of groups to STEWART PRIDDY, Department of Mathematics, Northwestern University, 2033 Sheridan Road, Evanston, IL 60208-2730; email: `priddy@math.nwu.edu`

Combinatorics and Lie theory to PHILIP J. HANLON, Department of Mathematics, University of Michigan, Ann Arbor, Michigan 48109-1003; email: `hanlon@math.lsa.umich.edu`

Complex analysis and complex geometry to DANIEL M. BURNS, Department of Mathematics, University of Michigan, Ann Arbor, MI 48109-1003; email: `dburns@math.lsa.umich.edu`

***Differential geometry and global analysis** to CHUU-LIAN TERNG, Department of Mathematics, Northeastern University, Huntington Avenue, Boston, MA 02115-5096; email: `terng@neu.edu`

***Dynamical systems and ergodic theory** to ROBERT F. WILLIAMS, Department of Mathematics, University of Texas, Austin, Texas 78712-1082; email: `bob@math.utexas.edu`

Geometric topology, knot theory, hyperbolic geometry, and general topoogy to JOHN LUECKE, Department of Mathematics, University of Texas, Austin, TX 78712-1082; email: `luecke@math.utexas.edu`

Harmonic analysis, representation theory, and Lie theory to ROBERT J. STANTON, Department of Mathematics, The Ohio State University, 231 West 18th Avenue, Columbus, OH 43210-1174; email: `stanton@math.ohio-state.edu`

***Logic** to THEODORE SLAMAN, Department of Mathematics, University of California, Berkeley, CA 94720-3840; email: `slaman@math.berkeley.edu`

Number theory to MICHAEL J. LARSEN, Department of Mathematics, Indiana University, Bloomington, IN 47405; email: `larsen@math.indiana.edu`

Operator algebras and functional analysis to BRUCE E. BLACKADAR, Department of Mathematics, University of Nevada, Reno, NV 89557; email: `bruceb@math.unr.edu`

***Ordinary differential equations, partial differential equations, and applied mathematics** to PETER W. BATES, Department of Mathematics, Brigham Young University, 292 TMCB, Provo, UT 84602-1001; email: `peter@math.byu.edu`

***Partial differential equations and applied mathematics** to BARBARA LEE KEYFITZ, Department of Mathematics, University of Houston, 4800 Calhoun Road, Houston, TX 77204-3476; email: `keyfitz@uh.edu`

***Probability and statistics** to KRZYSZTOF BURDZY, Department of Mathematics, University of Washington, Box 354350, Seattle, Washington 98195-4350; email: `burdzy@math.washington.edu`

***Real and harmonic analysis and geometric partial differential equations** to WILLIAM BECKNER, Department of Mathematics, University of Texas, Austin, TX 78712-1082; email: `beckner@math.utexas.edu`

All other communications to the editors should be addressed to the Managing Editor, WILLIAM BECKNER, Department of Mathematics, University of Texas, Austin, TX 78712-1082; email: `beckner@math.utexas.edu`.